Advances in Intelligent Systems and Computing

Volume 770

Series editor

Janusz Kacprzyk, Polish Academy of Sciences, Warsaw, Poland
e-mail: kacprzyk@ibspan.waw.pl

The series "Advances in Intelligent Systems and Computing" contains publications on theory, applications, and design methods of Intelligent Systems and Intelligent Computing. Virtually all disciplines such as engineering, natural sciences, computer and information science, ICT, economics, business, e-commerce, environment, healthcare, life science are covered. The list of topics spans all the areas of modern intelligent systems and computing such as: computational intelligence, soft computing including neural networks, fuzzy systems, evolutionary computing and the fusion of these paradigms, social intelligence, ambient intelligence, computational neuroscience, artificial life, virtual worlds and society, cognitive science and systems, Perception and Vision, DNA and immune based systems, self-organizing and adaptive systems, e-Learning and teaching, human-centered and human-centric computing, recommender systems, intelligent control, robotics and mechatronics including human-machine teaming, knowledge-based paradigms, learning paradigms, machine ethics, intelligent data analysis, knowledge management, intelligent agents, intelligent decision making and support, intelligent network security, trust management, interactive entertainment, Web intelligence and multimedia.

The publications within "Advances in Intelligent Systems and Computing" are primarily proceedings of important conferences, symposia and congresses. They cover significant recent developments in the field, both of a foundational and applicable character. An important characteristic feature of the series is the short publication time and world-wide distribution. This permits a rapid and broad dissemination of research results.

More information about this series at http://www.springer.com/series/11156

Wookey Lee · Carson K. Leung
Editors

Big Data Applications and Services 2017

The 4th International Conference on Big Data Applications and Services

Springer

Editors
Wookey Lee
Department of Industrial Engineering
Inha University
Incheon
South Korea

Carson K. Leung
Department of Computer Science
University of Manitoba
Winnipeg, MB
Canada

ISSN 2194-5357 ISSN 2194-5365 (electronic)
Advances in Intelligent Systems and Computing
ISBN 978-981-13-0694-5 ISBN 978-981-13-0695-2 (eBook)
https://doi.org/10.1007/978-981-13-0695-2

Library of Congress Control Number: 2018942520

Printed on acid-free paper

This Springer imprint is published by the registered company Springer Nature Singapore Pte Ltd.
The registered company address is: 152 Beach Road, #21-01/04 Gateway East, Singapore 189721, Singapore

Foreword

It is a great pleasure to invite you to the Fourth International Conference on Big Data Applications and Services (BigDAS 2017), which will be held in Tashkent, Uzbekistan on August 15–18, 2017. BIGDAS 2017 is hosted by National University of Uzbekistan and Korea Big Data Service Society.

The main topic of BigDAS 2017 is “Power to Change the World, Big Data”. The Big Data has become a core technology to provide innovative solutions in the many fields such as health care, manufacturing, social life, etc. The aim of BigDAS 2017 is to present the innovative results, encourage academic and industrial interaction, and promote collaborative research in Big Data and digital information management worldwide. We expect that publications of our conference will be a cornerstone for the further related research and technology improvements in the field of Big Data.

For BigDAS 2017, we accepted many high-quality papers, which will be presented in oral and poster sessions. The organizing committee of BigDAS 2017 has prepared 11 technical sessions. The topics of technical sessions are

- Big Data Models and Algorithms,
- Big Data Architectures,
- Big Data Applications,
- Big Data Mining and Analysis,
- Big Data Security,
- Big Data Visualization,
- Social Network Analysis,
- Internet of Things (IoT) and Health care,
- Big Data in Industry,
- Big Data in Government, and
- Big Data in Business.

We hope you will find the results presented during the conference useful and inspiring for your future research.

We would like to express our sincere thanks to invited speakers, organizing committee, and authors for their valuable contributions to the conference.

Suwon, South Korea
Joo-Yeon Lee
General Co-chair of BigDAS 2017
Ajou University

Tashkent, Uzbekistan
August 2017
Avazjon Marakhimov
General Co-chair of BigDAS 2017
National University of Uzbekistan

Preface

The International Conference on Big Data Applications and Services (BigDAS) aims to address the need of the academic community and industry about Big Data. It encourages academic and industrial interaction and promotes collaborative research in Big Data applications and services by bringing together academics, government, and industry professionals to discuss recent progress and challenges in Big Data applications and services. Moreover, BigDAS also serves as a platform for theoreticians and practitioners to exchange their original research ideas on academic or industrial aspects of Big Data applications and services, present their new findings or innovative results on theoretical or practical aspects of Big Data, share their experiences on integrating new technologies into products and applications, discuss their work on performing Big Data applications and services in real-life situations, describe their development and operations of challenging Big Data related systems, and identify unsolved challenges.

Since the First International Conference on Big Data Applications and Services (1st BigDAS 2015), three BigDAS conferences have been held in the following venues:

- **1st BigDAS 2015**: Seogwipo KAL Hotel, Jeju Island, South Korea, on October 20–23, 2015;
- **2nd BigDAS 2016**: Korea Software HRD Center, Phnom Penh, Cambodia, on January 22–27, 2016; and
- **3rd BigDAS-L 2016**: Landmark Mekong Riverside Hotel, Vientiane, Laos, on December 20–23, 2016.

With their success, the Fourth International Conference on Big Data Applications and Services (4th BigDAS 2017) is held in the following venues:

- **4th BigDAS 2017**: National University of Uzbekistan, Tashkent, Uzbekistan, on August 15–18, 2017.

The BigDAS 2017 was held in Tashkent, Uzbekistan (which is located in the center of Great Silk Road). The conference was hosted by (i) The Korea Big Data Service Society and (ii) National University of Uzbekistan. It was organized by (i) Bigdata

Research Institute, Chungbuk National University; (ii) New Industrial Convergence R&D Center, Ajou University; (iii) Inha University in Tashkent; and (iv) Korea China Yeouido Leaders Forum. It was sponsored by (i) Electronics and Telecommunications Research Institute (ETRI); (ii) UNISEM Co., Ltd.; and (iii) WIZCORE, Inc. The program consists of the following events:

- Four keynote speeches on big data in industry and government:
  - "The Fourth Industrial Revolution and New Industry Development Directions" by Marn-ki Jeong (ex-First Vice Minister of the Ministry of Trade, Industry and Energy, South Korea)
  - "Prediction Markets a Computational Mechanism for Aggregating Information" by Sarvar Abdullaev (Inha University in Tashkent, Uzbekistan)
  - "Big Data Approach and Challenges in Government" by Myoung Hee Kim (President of National Computing and Information Service (NCIS), Ministry of Interior, South Korea)
  - "Initiation of an Innovation Ecosystem: Uzbekistan's Road Towards Building Silicon Valley" by Bokhodir Ayupov (Ministry of Information Technologies and Communications, Uzbekistan)
- eight technical sessions with 54 regular and short paper presentations on the following topics: (i) Big Data models and algorithms, (ii) Big Data architectures, (iii) Big Data applications, (iv) Big Data mining and analysis, (v) Big Data security, (vi) Big Data visualization, (vii) social network analysis, and (viii) Internet of things (IoT) and health care;
- three poster sessions with eight poster paper presentations on the following topics: (i) Big Data applications, (ii) Big Data in industry, and (iii) Big Data in business; and
- two workshops: (i) E-government and Big Data, and (ii) Big Data composition.

For BigDAS 2017, we have recruited many international experts in Big Data applications and services to join our team of international committee. As a result, our Committee consists of professionals from different parts of the world such as Bangladesh, Canada, China, Japan, South Korea, and Uzbekistan. This committee has done an excellent job in organizing the conference and selecting a collection of 54 high-quality regular and short papers, as well as 8 poster papers. Among these papers, a small number of them were selected and revised to be included in the current volume.

BigDAS 2017 would not have been possible without the help and effort of many people and organizations. We thank BigDAS 2017 Organizing Committee members, especially Honorary Co-chairs (S. H. Han and N. Yusupbekov), General Co-chairs (J.-Y. Lee and A. R. Marakhimov), and Organizing Co-chairs (K.-H. Yoo, W.-S. Cho, and Y.-S. Kim) for their valuable advice and suggestions toward the conference. We are grateful to other members for their professionalism and dedication in different aspects of this conference, including the selection of papers presented at the conference and the further selection of papers published in the current volume. We also express our thanks to authors and non-author participants of this conference. We also thank A. Nasridinov for local arrangement. Last but not least, we thank the staff at Springer for their help in publishing the current volume.

Incheon, South Korea
Wookey Lee
Program Co-chair of BigDAS 2017

Winnipeg, Canada
Carson K. Leung
Program Co-chair of BigDAS 2017

August 2017

Organization

Honorary Co-chairs

Sun Hwa Han, KISTI, South Korea
Nadirbek Yusupbekov, Tashkent State Technical University, Uzbekistan

General Co-chairs

Joo-Yeon Lee, Ajou University, South Korea
Avazjon R. Marakhimov, National University of Uzbekistan, Uzbekistan

Program Co-chairs

Wookey Lee, Inha University, South Korea
Carson K. Leung, University of Manitoba, Canada

Organizing Co-chairs

Kwan-Hee Yoo, Chungbuk National University, South Korea
Wan-Sup Cho, Chungbuk National University, South Korea
Yoo-Sung Kim, Inha University, South Korea

Local Arrangement Co-chairs

Aziz Nasridinov, Chungbuk National University, South Korea
Sarvar Abdullaev, Inha University in Tashkent, Uzbekistan
Azizbek Marakhimov, Dongguk University, South Korea

Publicity Co-chairs

Sang-Hyun Choi, Chungbuk National University, South Korea
Seon-Phill Jeong, United International College, China

Web Co-chairs

Young-Koo Lee, Kyung Hee University, South Korea
Phyoung-Jung Kim, Chungbuk Provincial College, South Korea

Industry Co-chairs

Tae-Sung Kim, Chungbuk National University, South Korea
Soo-young Chae, National Security Research Institute (NSRI), South Korea

Industry Liaison Co-chairs

Duck-Keun Park, Wizcore, South Korea
Jung-Tae Kang, YURA, South Korea

Demo/Poster Co-chairs

Su-young Chi, ETRI, South Korea
Nakhoon Baek, Kyungpook National University, South Korea

Registration Co-chairs

Young-Ho Park, Sookmyung Women's University, South Korea
Yong-Ik Yoon, Sookmyung Women's University, South Korea

Tutorial Co-chairs

Jin-Ho Kim, Kangwon National University, South Korea
Supratip Ghose, UITS, Bangladesh

Special Session Co-chairs

Eun-mi Choi, Kookmin University, South Korea
Jin-young Choi, Ajou University, South Korea
Takeshi Kurata, AIST, Japan
Tianhong Pan, Jiangsu University, China
Masaki Aono, Toyohashi University of Technology, Japan

Sponsor Co-chairs

Seong Joon Yoo, Sejong University, South Korea
Jong-Seok Myong, Oracle Inc., South Korea

Secretariat

Jong-Min Yoon, KITECH, South Korea

Co-hosts

Korea Big Data Service Society

NATIONAL UNIVERSITY OF UZBEKISTAN

Organizers

Sponsors

Contents

3D Volume Visualization System Based on GPUs for Medical Big Data

Joseph Yoon, Oh-Soek Kown, Jae-Gyun Cheong, Keon-Kuk Park and Young-Bong Kim

Abstract With the appearance of computerization society, medical equipment has been closely coupling with computers. Images from medical equipment have become very useful basic data in diagnosing the state of a patent. These images have often obtained from medical computerized equipment such as X-ray system, Ultrasound system, Computerized Tomography (CT), Magnetic Resonance Imaging (MRI), and so on. For increasing the effective value of those images, we have to employ the preprocessing techniques such as a median filter, a Gaussian filter, and a mean filter. And also, we take the special algorithms such as a phase correction, a dispersion compensation, and a Hilbert transform. These additional techniques make it impossible to watch the images in a real time. Thus, we have to get a rapid processing of medical images for the real-time display of those images. In this research, we will propose a method to give a real-time display of images using multi-thread techniques and parallel processing algorithms. In a multi-thread technique, if two threads share the same resource, one thread may change the resource data during the processing of the other thread. We will design a method to avoid the collision in a resource usage right between threads.

Keywords Triple buffers · Dual GPU · Medical image processing · 3D volume

J. Yoon · O.-S. Kown · J.-G. Cheong · K.-K. Park · Y.-B. Kim (✉)
IT Convergence and Application Engineering, Pukyong National University,
Yongso-Ro. 45, Nam-Gu, Busan 48513, Korea
e-mail: ybkim@pknu.ac.kr

J. Yoon
e-mail: ricerich@gmail.com

O.-S. Kown
e-mail: inartistical@gmail.com

J.-G. Cheong
e-mail: jaegyuncheong@gmail.com

K.-K. Park
e-mail: pkk1113@naver.com

W. Lee and C. K. Leung (eds.), *Big Data Applications and Services 2017*,
Advances in Intelligent Systems and Computing 770,
https://doi.org/10.1007/978-981-13-0695-2_1

1 Introduction

OCT (Optical coherence tomography) is an established medical imaging technique that uses light to capture micrometer-resolution, three-dimensional images from within optical scattering. This is employed in diverse applications, including art conservation and diagnostic medicine, notably in ophthalmology and optometry where it can be used to obtain detailed images from within the retina. Recently, it has also begun to be used in interventional cardiology to help diagnose coronary artery disease [1]. Compared with Computed Tomography (CT), Magnetic Resonance Imaging (MRI), and Ultrasonography, it can provide high-resolution data and also observe the speed of hematocele (=blood flow). However, OCT data processing spends much time because it has to perform complex calculation from acquisition data of device to image of screen, and also in order to get better high-quality image, it needs to process more huge data and employ more image processing algorithms.

Due to these reasons, several researchers have approached in two ways. Firstly, as software method [2, 3], they focus on filtering to reduce noise like mean filter, Gaussian filter, median filter, and algorithm improvement of data process step to compensate data like phase correction, Hilbert transform, and dispersion compensation. Secondly, as hardware method, they research the improvement of calculation efficiency by multi-thread method on central processing unit (CPU)- or GPU based parallel computing [4, 5].

In this paper, in order to get high-quality image by software approach, we employ advanced filtering and multiple processing algorithm. And also, as hardware approach, we implement GPU based parallel computing instead of CPU based. We also reduce its calculation cost by GPU based parallel computing and we implement the distribute module divided into image processing using dual GPU and 3D volume visualization. In a multi-thread method, if shared-memory conflict and waiting-delay problems occur, we can solve them by employing the Triple buffer technique. Finally, we can obtain real-time 3D visualization image from OCT device.

2 Medical 3D Volume Visualization System

2.1 System Architecture

Our system configuration is shown in Fig. 1. This FD-OCT (Fourier Domain Optical Coherence Tomography) visualization system consists of Triple buffers in CPU and in dual GPU based volume rendering.

First of all, raw data stream is stored continuously from OCT device like CMOS (complementary metal-oxide-semiconductor) line scan camera to Buffer1 in CPU, and then it is transferred to GPU1 which calculates parallel FD-OCT processing. After filtering and algorithm processing in GPU1, it stores to Buffer2-1, Buffer2-2, or Buffer2-3 at the CPU as unit data possible to be rendered volume visualization.

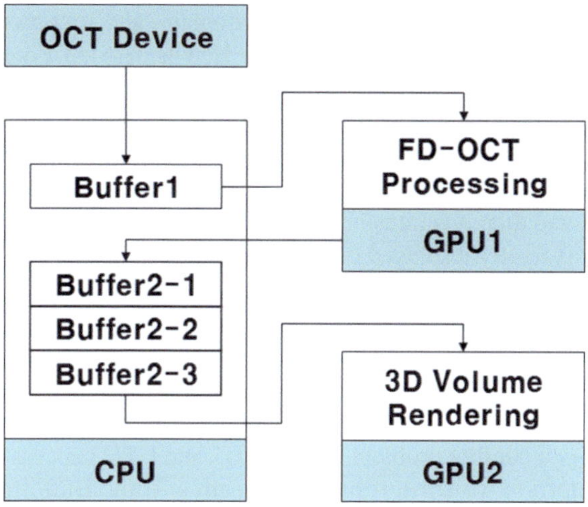

Fig. 1 System overview

When volume unit data is made, it moves to GPU2 which calculates parallel volume visualization, and finally it is displayed on the screen.

2.2 *FD-OCT Processing*

FD-OCT processing is implemented employing the Kang [5] algorithm to get clear OCT image as shown in Fig. 2. It is designed at GPU1 to have parallel calculation with guaranteed higher speed.

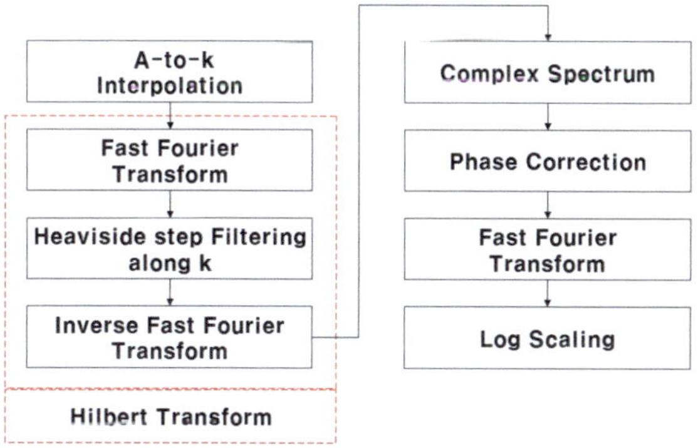

Fig. 2 FD-OCT processing

First of all, input data is interpolated by k-domain, and it is processed through Hilbert Transform (Fast Fourier Transform → Heaviside Step Filtering → Inverse Fast Fourier). Its result comes to be Complex Spectrum. Next step is Phase Correction to compensate spread of image by optical device adjustment and constant value setting. After it is processed with the Fast Fourier Transform again, the OCT frame image is obtained in the Log Scaling step. Its generated data is transferred to Triple buffers in CPU, and then it is transferred to GPU2 which calculates parallel volume visualization. It is finally displayed on the screen.

2.3 *Triple Buffering*

The proposed system has to share CPU memory because it utilizes dual GPU. However, there occurs a conflict problem when GPU1 and GPU2 access the CPU at the same time. Therefore, we solve it by implementing Triple Buffering as shown in Fig. 3.

As shown in Fig. 3, size of time slot 1 at GPU1 is unit size of FD-OCT processing, that means available data which can be used for volume visualization. Generated volume data after GPU1 processing is written at one of the triple buffers if it is not read by GPU2. At the same time, GPU2 reads volume data from CPU data written by GPU1, and then it displays volume visualization.

Size of time slot 1 at GPU2 is unit size of volume rendering, it is processed repeatedly because it waits until next volume data is generated by writing unit data from GPU1 to CPU, and it is also for keeping rendering. For example, as shown in Fig. 3, in order to generate possibly visualizable volume data in GPU1 will be used in GPU2, FD-OCT processing operates with six units, and its data is written alternately

Fig. 3 Dual GPU based triple buffering

on one of the triple buffers at the CPU, and finally it performs volume visualization five times using its written data at GPU2.

3 Experiments and Results

We tested the rendering performance of two systems as illustrated in Fig. 4. One is previously ① single GPU based system, the other is ② our dual GPU based and Triple Buffering system. In single GPU based system, it performs FD-OCT processing first, and sequentially renders 3D volume data. On the other hand, in our dual GPU based and Triple Buffering system, rendering performance is faster than previous system at second time slot. This is because two tasks perform, respectively, which are FD-OCT processing and 3D volume visualization.

As shown in Fig. 4, time slot (ex:1-a) of FD-OCT processing is unit data for 2D image processing. When k (=depth) numbers of 2D image are combined, it becomes one of 3D volume available data (ex:1-a,1-b,…,1-k). This generated data is displayed repeatedly in ② our system until next data generation is completed. For example, time slot 1 of volume rendering in ② our system is displayed three times repeatedly.

In our experiment, we used a 12-bit dual-line CMOS line scan camera (Sprint spL2048-140 km, Basler AG, Germany), data acquisition (DAQ) Card, frame grabber (NI PCIe-1429, National Instruments, Austin, TX). The PC system configuration is

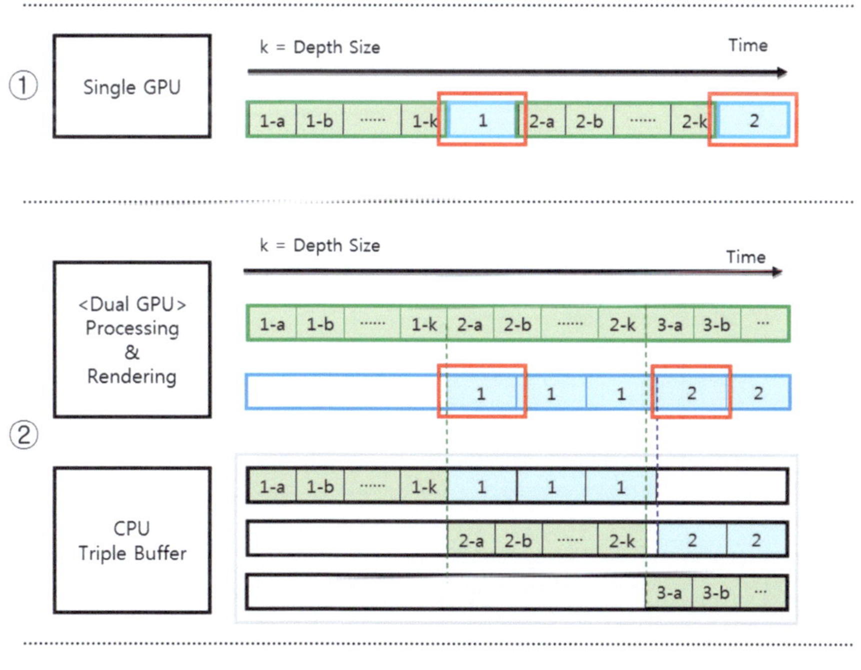

Fig. 4 Compare with 1 GPU-based system and our system

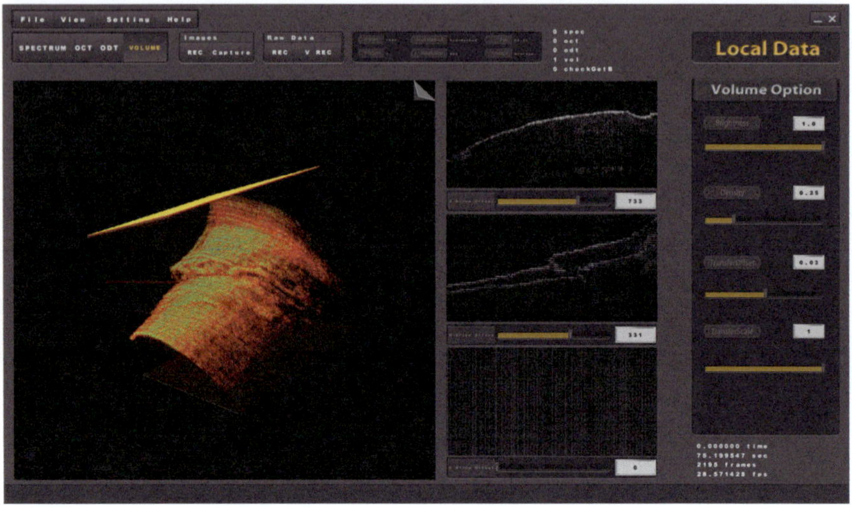

Fig. 5 Result with 3D volume visualization

Intel Core™ i7- 3970X CPU, 8 GB Memory, NVIDA GeForce GTX 760 (Memory 4 GB, CUDA Core 1024), and OS (operating system) is Window 7 Professional K 64 bit version.

As shown in Fig. 5, it is a result of 3D volume visualization by the implemented system. As shown in Fig. 6, it represents rendering speed using the 32 or 64 slices of 2D image (1024 × 1024), and also it compares ① single GPU based system with ② dual GPU and Triple buffers system.

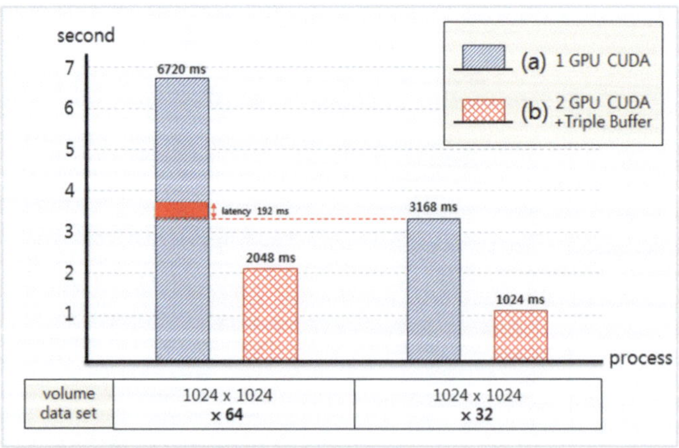

Fig. 6 Compare with 1 GPU-based system and our 2 GPU and Triple buffers system

Our dual GPU and Triple buffers system is three times faster than the previous single GPU-based system.

4 Conclusion

In this paper, we have designed and implemented 3D volume visualization system. Thus, we have proposed dual GPU and Triple Buffering techniques. In result, we have obtained high-quality display with improved rendering speed. We expect our system to be utilized in actual medical image processing.

References

1. Sebastian, M., Birgit, S., Mette, M. Thomas, M.J., Peter, E.A.: Optical Coherence Tomography—current Technology and Applications in Clinical and Biomedical Research. Anal. Bioanal. Chem. pp. 2699–2720 (2011)
2. Hossen, A.: A New Fast Approximate Hilbert Transform with Different Applications. IIUM Eng. J. pp. 21–27 (2001)
3. Shuichi, M., Tapio, F., Yoshiaki, Y.: Full-range, High-speed, High-resolution 1-μm Spectral-Domain Optical Coherence Tomography using BM-scan for Volumetric Imaging of The Human Posterior Eye. Opt. Express. pp. 8406–8420 (2008)
4. Yong, H., Xuan, L., Jin, U.K.: Real-Time 3D and 4D Fourier Domain Doppler Optical Coherence Tomography based on Dual Graphics Processing Units. Biomed. Opt. Express. pp. 2162–2174 (2012)
5. Kang, Z., Jin, U.K.: Graphics Processing Unit-based Ultrahigh Seed Real-Time Fourier Domain Optical Coherence Tomography. Sel. Top. Quantum. Electr. pp. 1270–1279 (2012)

A Fire Flame Simulation Scheme with Massively Parallel Processing

Byeonguk Im and Nakhoon Baek

Abstract Simulating natural phenomena is one of the most important areas in computer graphics area. As one of the natural phenomena simulations, we focused on the simulation of fire flames. Previous fire flame simulations mainly focused on the simulation sequences. At this time, those methods are not suitable for modern computer graphics and massively parallel processing architectures. In this work, we present a prototype implementation of the fire flame simulation system, based on the Compute Unified Device Architecture (CUDA) and Open Graphics Library (OpenGL). Our system shows highly efficient execution of those simulations, to show the real-time fire flame simulations.

Keywords Fire flame · Simulation · Massively parallel

1 Introduction

In the field of computer simulation, the natural phenomena simulation is one of the important applications [1, 2]. Among its various simulation targets, we focused on the simulation of fire flames. Historically, many different methods applied to the fire flame simulation, including particle systems, fluid dynamics approaches, spatial grid acceleration techniques, and others [3, 4].

In this work, we will use a particle system, which uses a large number of particles to simulate the motions of various kinds of gases, fluids, and small objects. The

B. Im · N. Baek (✉)
School of Computer Science and Engineering, Kyungpook National University, Daegu 41566, Republic of Korea
e-mail: oceancru@gmail.com

N. Baek
Software Technology Research Center, Kyungpook National University, Daegu 41566, Republic of Korea

N. Baek
Dassomey.Com Inc, Daegu 41566, Republic of Korea

W. Lee and C. K. Leung (eds.), *Big Data Applications and Services 2017*,
Advances in Intelligent Systems and Computing 770,
https://doi.org/10.1007/978-981-13-0695-2_2

particle simulation systems are widely used for various computer games, computer animations, graphics simulations, and other fields, since it is relatively simple and efficient to show visually acceptable natural phenomena.

In contrast, the heavy computation due to the particle interactions and the extremely large number of particles is one of their weak points. There were many trials to accelerate the particle system with parallel processing. In this work, we also use the massively parallel processing system of CUDA [5] as the acceleration framework for the simulation of fire flames. It is a new particle system design with CUDA-based processing and targeting the fire flame simulation. We present detailed descriptions on our prototype design, in the following sections.

2 Previous Works

2.1 Particle System

A particle system uses a large number of very small elements, to simulate some kinds of physical phenomena. It can generate some natural phenomena, including fire, explosion, smoke, and others, even with relatively small amount of computations. It is also easy to reproduce the visually plausible rendering sequences.

There are many implementations of particle systems. In this work, we focus on the fire flame simulation, as a good example of complex particle system with extremely large number of particles. In contrast, there are relatively small number of particle system implementations, on the massively parallel architectures including CUDA [5] and DirectCompute [6].

2.2 Smoothed-Particle Hydrodynamics

Smoothed-particle hydrodynamics is a computational model to simulate the dynamics of continuum media [7]. In the field of computer graphics, it is widely used for fluid simulation. This method has several benefits over the traditional methods. Since it uses the particle systems as the underlying framework, it is easy to maintain the physical properties such as mass, acceleration, velocity, and temperature. In contrast to other techniques, this smoothed-particle hydrodynamics system is free from any loss of physical measurements.

In this work, we plan to implement a particle system for fire flame simulation, based on the massively parallel architecture for the large amount of particles. The smoothed-particle hydrodynamics calculations also applied to the physical simulations of the particles. The details follow in the following sections.

3 Design of Massively Parallel System

In this work, we aim to reproduce the fire flames, through exactly simulating the massive amount of particle interactions. Every particle interacts with other particles and environmental objects. It will be finally disappeared after the complete combustion. To simulate the whole fire flame sequences, we will use the following modules to construct the massively parallel particle simulation system:

- particle emission
- combustion
- grid indexing
- sorting the particles
- physical simulation
- graphics rendering.

We will present the detailed descriptions on these modules.

3.1 Particle Emission

As the first step, we generate a set of new particles. All the particles emitted from the pre-specified regions in the 3D space. The system keeps track of all the particles, including their total numbers, their states, current positions, velocities, and others. All the properties are kept in the current state table.

3.2 Combustion

Our system targets the simulation of fire flames. Therefore, the particles should show the combustion process. We can simulate all the combustion processes through changing the color of the particles. Especially, we will change the color of particles with respect to the simulated temperature of the particles, finally to show the dramatic effects of the fire flame.

Figure 1 shows the natural color changes during natural combustions. We can simulate the realistic particles using the color levels shown here. The colors will change according to the elapsed time, current combustion status, the original burning substances, and others.

3.3 Grid Indexing

In our system, the physical forces applied to particles are calculated using smoothed-particle hydrodynamics theories. In this calculation, each particle should interact with all other particles located in the specified radius. Therefore, in a naïve manner, we

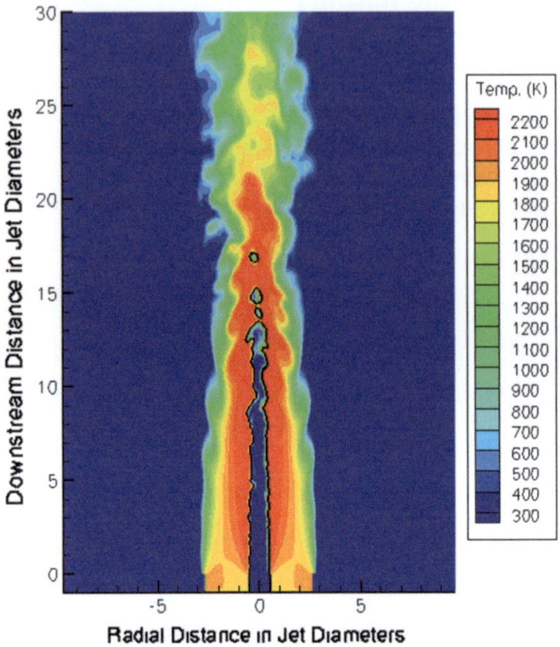

Fig. 1 Color changes during natural combustions

need to check all the distances between all the particles, to check whether a pair of particles has the distance less than the specified radius or not. All these calculations need heavy computations, and thus, we introduce an acceleration technique of grid-based approaches.

We divide the whole domain into a 3D regular grid structure. Figure 2 shows the two-dimensional version of our grid approach. Every grid will maintain its own list of active particles in that grid, and will check all the entrance and exit of all the particles. Even with these maintenance costs, the overall calculation is accelerated to achieve real-time simulations.

3.4 Sorting the Particles

Before applying the physical simulations to the particles, we need to arrange the order of particle simulation, for each grid, and also, for all the global space. Actually the particles are arranged several times, according to a certain property.

In typical massively parallel processing architectures including CUDA [5] and DirectCompute [6], the sorting operations are one of the most difficult ones for the accelerated implementation. In our case, we should handle millions of particles, and therefore, we should carefully handle the internal data caches of the architecture. More precisely, we used a variation of the bi-tonic sorting algorithm [8].

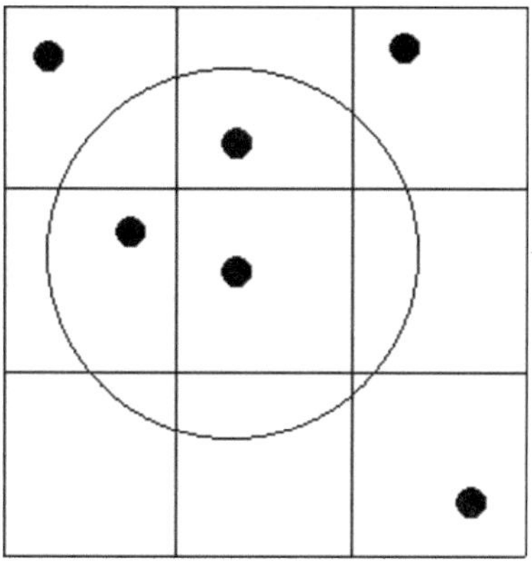

Fig. 2 The 2D representation of our regular grid

3.5 Physical Simulation

The physical simulation of the particles is actually the process of calculating all the physical properties of the particles. They include force, acceleration, velocity, temperature, and others. We used the smoothed-particle hydrodynamics theories to get all these physical properties. The calculation itself can be achieved in a massively parallel manner, with our implementations.

3.6 Graphics Rendering

The final result of our simulation is the position and the color of every active particle in the whole domain. These positions and colors can be simulated mapped as the positions and colors of a very large size point cloud. We use the 3D graphics library of OpenGL [9] to draw this point cloud. Additionally, for full speed-up of this rendering process, we use the shared texture area between CUDA and OpenGL, to accelerate both of the physical simulation and the graphics rendering process.

4 Experimental Results

Our experimental system consists of the following components:

- Intel Core i5-6600 CPU, 3.30 GHz
- NVIDIA GeForce GTX 1060, with 6 GByte Graphics Memory
- 8 GByte Main Memory
- Windows 8.1 64 bit version.

Fig. 3 Animation sequences from our fire flame simulation

Integrating all the software modules explained in the previous section, we executed this prototype system and get the results in Fig. 3. The actual execution generates a real-time animation of fire flames, and Fig. 3 shows some screen shots from the animation sequences.

5 Conclusion

In this work, we have implemented a prototype system of fire flame simulation, with the massively parallel architecture support. Specifically, we used CUDA as the massively parallel computation framework. Basically, our work has the benefits of using massively parallel computation for an extremely large number of particles. In previous works on fire flame simulation, they focused on the acceleration of the sequential computation. In contrast, our work focused on the throughput of the large number of data elements, and we show that even fire flame simulations can be accelerated by the massively parallel architecture.

In near future, our prototype system will be updated with some modifications. They include the enhancement on the grid structures and their processing. We also plan to add more kinds of particle properties. For example, the smokes and fire flakes can also be simulated with our particle simulation system.

Acknowledgements This research was supported by Basic Science Research Program through the National Research Foundation of Korea (NRF) funded by the Ministry of Education, Science and Technology (Grant No. 2016R1D1A3B03935488).

References

1. Fedkiw, R.: Simulating Natural Phenomena for Computer Graphics. (2002).
2. Hughes J.F., et.al.: Computer Graphics: Principles and Practices, 3rd Ed., (2013).
3. Lyes, T.S. and Hawick K.A.: Fire and Flame Simulation using Particle Systems and Graphical Processing Units, Proceedings of the International Conference on Modeling, Simulation and Visualization Methods (MSV); Athens, (2013).
4. Rodal, K.E.S. and Storli G.: Physically Based Simulation and Visualization of Fire in Real-Time using the GPU, Dept. of Computer and Information Science, Norwegian University of Science and Technology, (2006).
5. NVIDIA, CUDA C Programming Guide, NVIDIA, (2017).
6. NVIDIA, DirectCompute Programming Guide, NVIDIA, (2010).
7. Vorobyev, A.: A Smoothed Particle Hydrodynamics Method for the Simulation of Centralized Sloshing Experiments, KIT Scientific Publishing, (2013).
8. Peters H., Schulz-Hildebrandt O., Luttenberger N.: Fast In-Place Sorting with CUDA Based on Bitonic Sort. Parallel Processing and Applied Mathematics. PPAM 2009. Lecture Notes in Computer Science, vol 6067. Springer, Berlin, Heidelberg, (2010).
9. Segal, M., Akeley, K.: The OpenGL Graphics System: A Specification, Version 4.5 (Core Profile). Khronos Group, (2016).

A Framework for Calculating Damages of Personal Information Leakage Accidents

Dong-Hee Lee and Tae-Sung Kim

Abstract Victims of personal information leakage accidents have not been adequately compensated, and large-scale leakage accidents are occurring, which is becoming a serious social problem. In this study, the causes of the personal information leakage accidents that continue to occur are due to insufficient compensations for victims and lower costs to be borne by the companies. This resulted in the loss of the enterprise's reason for investing in information security, which in turn led to an average low security level. Therefore, we analyze the cases of personal information leakage accidents in Korea and present a framework for calculating damages.

Keywords Personal information leakage · Data breach · Damages · Framework

1 Introduction

Modern people have used personal information to authenticate themselves in non-face-to-face situations such as work, shopping, banking, etc., through online. Therefore, the personal information represents himself, and the risk of identity theft by others is also increased. However, more than 10 million personal information leakages occurred in Korea in 2008, and large information leakage accidents are continuously occurring after that. Therefore, in this study, it was decided that the appropriate compensation system was needed as a measure to prevent continuous information leakage accident. Accordingly, we propose a framework for calculating damages in Korea, and this gives companies some insights into the direction of effective security

D.-H. Lee
Department of Information Security Management and Cybersecurity Economics Research Institute, Chungbuk National University, Cheongju, Korea
e-mail: iluy3082@gmail.com

T.-S. Kim (✉)
Department of Management Information Systems and Cybersecurity Economics Research Institute, Chungbuk National University, Cheongju, Korea
e-mail: kimts@cbnu.ac.kr

W. Lee and C. K. Leung (eds.), *Big Data Applications and Services 2017*,
Advances in Intelligent Systems and Computing 770,
https://doi.org/10.1007/978-981-13-0695-2_3

enhancement, while also providing what needs to be supplemented for victim relief or accident prevention.

2 Personal Information and Damages

2.1 Personal Information

According to the Act on promotion of information and communications network utilization and information protection, etc., personal information is information about an individual who is alive and includes information such as code, character, voice, sound, and image that can identify the individual by name, resident registration number, even if you cannot recognize an individual but can easily identify him with other information, you can define it as personal information [1]. In other words, if a person can be identified by combining physical information such as a physical characteristic for identifying each individual, a number assigned to an individual in a society, an International Mobile Equipment Identity (IMEI) number of a smartphone purchased by an individual, and a media access control (MAC) address of a local area network (LAN) card, it can be called personal information [2]. On the other hand, as an important feature of personal information, personal information is included as a type of goods and has value as things with physical reality [3]. Therefore, personal information is targeted by malicious objects for monetary gain. Leaked personal information can be used to sell a business as it is, to resell it in combination with other information, or to intimidate a company that has leaked information.

2.2 Personal Information Leakage and Damages

According to the Supreme Court's case, personal information leakage is a state in which personal information is out of the control of an information and communication service provider and a third party becomes aware of the contents [4]. This is a matter related to the right of self-determination of personal information, which means that the right of the information subject to self-determination of the disclosure and use of personal information is infringed [5]. At this time, the types of the personal information processor who are liable for personal information leakage damages are as follows. First, when personal information is leaked due to the negligence of the personal information processor. Second, when personal information is leaked due to the combination of the negligence of the personal information processor and illegal acts caused by intentional hacking by a third party. Third, in case of leakage of personal information due to intention of personal information processor's employee or trustee. Fourth, when the personal information processor provides or transfers personal information to a third party without obtaining consent from the information subject. Fifth, a case where the personal information processor collects information without consent of the information subject, uses it differently from the purpose agreed, or does not fulfill the request even if the information subject is withdrawn, viewed, or corrected [6]. However, in case of damages in the case of Korea, it is mainly the concept of compensation for the mental damage of the information subject due to leakage of personal information, and if the company proves that it does not violate the technical and administrative protection measures, there are many cases in which compensation is not made. In addition, even if it appears that there is a vulnerability to leakage of personal information due to technical and managerial problems of the company, the fact alone does not constitute a compensation for damages. Similarly, the second damage such as monetary damage if the probability is unclear and the damage cannot be proven, the claimant of the leaked victim is not acknowledged [7]. A large leaked accident followed the personal information leaked accident in 2014, but there was no proper compensation for victims and future security measures. Accordingly, the punitive damages system and the statutory damages system were introduced through amendment of the Personal Information Protection Act in July 2015 [8]. The punitive damages system is punishable by compensating for more than damages if the perpetrator's actions are intentional, malicious, or antisocial [9]. The Personal Information Protection Act stipulates that personal information should be repaid up to three times the amount of damages in the event of loss of information, such as loss, theft, or leakage, due to intentional or grave error of the personal information processor. However, there is a problem that it is difficult for the victim to prove that the victim is not punishing the perpetrator because the financial information does not occur to the information subject only by the leaked fact, and that even if the financial damages occur, the damage is caused by the leakage of the personal information. In order to compensate for this, if the legal damages system which is difficult to substantiate the damage occurred to the victim, he can claim damages within the range of 3 million won or less instead of the actual damages. In this case, the information communication service provider, etc., should prove that there is no intention or negligence [8]. As a result of the introduction of these systems, it is expected that the rights of information subjects will be enhanced as the effect of punitive damages and the responsibility of proving damage are shifted from information subjects to information communication service providers. However, due to the prolonged length of the litigation period, victims abandon their lawsuits and only a few victims participating in the litigation are subject to damages, so few victims are compensated for the actual accident. The compensation amount is still insignificant. Therefore, the company has a problem in that the cost of litigation and compensation for damages are so much lower than the cost of increasing the security investment right away in terms of cost advantage, so that the company chooses to repair the accident when an accident occurs without increasing the budget for security. Therefore, it is necessary to establish a criterion for the amount of damages to be appropriated to compensate victims. Also, it can be used as a criterion for determining the reasonable level of security by using as a criterion for establishing the appropriate amount of investment to prevent information leakage accidents. Therefore, based on the precedents and related studies of personal information leakage accidents, a framework for calculating damages for personal information leakage accidents is presented.

2.3 Framework for Calculating Damages for Personal Information Leakage Accidents

As a precedent study on the calculation of damages for personal information leakage accidents, there is a study on the standard of damages created by the Personal Information Dispute Mediation Committee [10]. The study presupposed that the damage caused by the personal information leakage accident was the psychological damage of the spill victims. In addition, the criteria for the calculation of damages for personal information leaked accidents are presented by referring to examples of US and Japan, methods of calculating damages due to various disasters in Korea, and calculation methods of calculating alienation in other fields. The criteria for the calculation include the period of infringement, the frequency of infringement, the sensitivity of infringement information, the scope of infringement information exposure, the victim's fault, the benefits gained through infringement, the motives and purpose of infringement, the disadvantages in social life, the social responsibility of the perpetrator, and the occurrence of follow-up loss are classified as 1–5, respectively, and the sum of the corresponding grade is summed up to provide a standard for calculating the amount of damages. On the other hand, there are researches on the criteria of calculating the amount of damages in consideration of the importance of personal information, the degree of damage, and the negligence of the corporate [11]. Kwon Hong et al. [12] conducted a study on CVM (Contingent Valuation Methods) to determine the average payment amount to prevent personal information leakage. On the other hand, the Japan Network Security Association (JNSA) proposed the Joint Strategic Needs Assessment (JSNA) JNSA Damage Operation (JO) Model considering the value of outgoing personal information, organizational social responsibility, and corporate accident response [13].

However, each criterion presented in the previous studies has a limitation that it is difficult to grasp the relative importance of the criteria. Therefore, there is a problem that it is necessary to decide which criteria should be focused first in order to prevent leakage of personal information from the viewpoint of the company and minimize the responsibility of accidents in case of information leakage. So, in this study, factors that are considered in the calculation of the damages based on previous studies and Korean precedents are classified according to importance. In addition, a new damages calculating framework is proposed as shown in Fig. 1 by adding some criteria that affect the amount of damages. A detailed description of the framework follows. First, if neither 1 nor 2 does exist, no damages will be incurred. If the damages are incurred, the size of the damages is determined by taking into account the numbers 3 and 4. The first criterion is the case where the information is spread

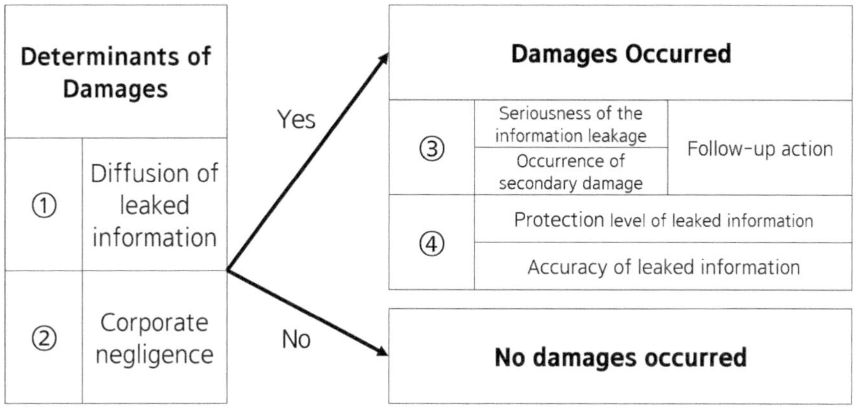

Fig. 1 Framework for calculating damages of personal information leakage accidents

to third parties and the information subject cannot control it. The second criterion is that the cause of the leakage was caused by the fault of the company.

3 Conclusion and Implications

As mentioned in the Supreme Court's ruling on the auction accident, complete security is hard to anticipate due to the fundamental nature of the Internet. In many cases, however, personal information has been leaked because of the level of security that violates regulations or fails to meet recommendations. Even in this case, the appropriate punishment is not imposed for the company and does not take responsibility for the victim. As this behavior continues, companies are no longer aiming to safely manage customer information, and it is highly likely that security policies will be developed in a favorable direction by comparing the costs of security investments and leaked accidents. If this condition persists, the overall security level of the enterprises may be lowered and the information leakage may become more vulnerable. Therefore, if the company's negligence is revealed, it is necessary to enforce an appropriate level of security and to make damages to the information subject with strong punishment. In this study, we present a framework for calculating damages for personal information leakage accidents by analyzing previous studies and case studies. The implications of this study are as follows: First, the criteria for liability for damages in the event of a leakage accident are suggested. In addition, by presenting the factors that determine the size of the damages, companies will be able to utilize security investment decisions when they want to invest in security. Second, from the viewpoint of the victims of the information leakage accident, this study will help to understand the criteria of damages more easily. Lastly, if future studies can provide concrete damages for each criterion, it can be used as a clue to dramatically shorten the length of litigation. It is hoped that this will contribute to the relief of victims suffered from information leakage accidents.

Acknowledgements This research was supported by the MSIT, Korea, under the "Employment Contract based Master's Degree Program for Information Security" supervised by the KISA (H2101-17-1001) and Chungbuk National University Cybersecurity Economics Research Institute (CERI).

References

1. Korea Communications Commission.: Act on Promotion of Information and Communications Network Utilization and Information Protection, etc. (2017)
2. Ryu Seung-Hoon.: A claim of compensation according to personal information leaks. Law Review, 47, 125–146 (2016)
3. Korea Online Privacy Association.: Personal information value and Analysis of social cost due to infringement of personal information. Personal Information Protection Commission (2013)
4. Supreme court of Korea.: Sentenced 2011da24555 rulings (2014)
5. Constitutional court of Korea.: Sentenced 2003hun ma282, 425 rulings (2005)
6. Koh Hong-Seok.: Case study: The occurrence of mental damages that could be compensated for by alimony due to leakage of personal information-Target ruling: Supreme court of Korea 2012. 12. 26. Sentenced 2011 da 59834, 59841, 59858 rulings. BFL, 66(67–83) (2014)
7. Kwak Young-am.: A Study on Ruling on Accident of Personal Information Leak in Korea. Journal of Korean Electronic Commerce Research Association 15, 63–81 (2014)
8. Privacy Act (Act No. 14389). (2017) (in Korean)
9. PMG Knowledge Engine Institute.: Punitive Damage. current common sense dictionary, http://terms.naver.com/entry.nhn?docId=72502&cid=43667&categoryId=43667
10. Lee Chang-bum, Kim Bon-mi.: A Study on Personal Information Infringement Remedy and Compensation Standard. Personal Information Dispute Mediation Committee Research Report (2004)
11. Cha Geon-sang.: A Study on the Criteria to Estimate the Compensation from the Infringement of Personal Information. Doctoral thesis, Soongsil University (2011)
12. Kwon Hong, Lee Eun-Joo, Kim Tae-Sung, Jun Hyo-Jeong.: Estimating Compensation for Personal Information Infringement in Korea Using Contingent Valuation Methods. Journal of The Korea Institute of Information Security & Cryptology 22(2), 367–377 (2012)
13. NPO Japan network security association.: 情昱セキュリティインシデントに関する調査昱告書 春紙 (in Japanese) (2017)

A Hierarchical Structure for Representing 3D Respiration Organ Models

Eath Manith, Chan Park and Kwan-Hee Yoo

Abstract In this paper, we propose a visualizing method that facilitates the 3D model representation using menu with level categories. With the implementation of the organs modeling method, the 3D internal respiratory tracts are incredibly separated into the various organs of both upper and lower regions based on function or anatomy. These regions are applied to the real color texture and materials by using the 3D creation tools. Moreover, the composed category level structuring of the object model is defined using the tree structure. Based on the tree structure, a TreeView technique for representing a navigation control that displays a hierarchical collection of labeled items is developed. It has functions such as expand and hide, drag and drop, and remove and replace. Using computational modeling and visualizing the respiratory system can enhance the understanding on how the 3D organs are categorized and performed with multiple anatomies of the human internal organs.

Keywords Visualization system · 3D respiration organs · 3D model partition Category level structuring · 3D creation tool · TreeView technique

1 Introduction

In medical science, research advancement nowadays comes from realistic models of human organs. This advancement can be observed from a production of the 3-dimensional (3D) model from computed tomography data that includes simulated

E. Manith · K.-H. Yoo (✉)
Department of Computer Science, Chungbuk National University, Cheongju, South Korea
e-mail: khyoo@chungbuk.ac.kr

E. Manith
e-mail: eath.manith@gmail.com

C. Park
KIS, Seongnam, South Korea
e-mail: szell.park@gmail.com

W. Lee and C. K. Leung (eds.), *Big Data Applications and Services 2017*, Advances in Intelligent Systems and Computing 770,
https://doi.org/10.1007/978-981-13-0695-2_4

respiratory organs of oral cavity-pharynx and larynx, and seven generations of the airways [1]. The respiratory system is a collection of internal organs and tissues that provides oxygen to the body's cells of the human being. To fully understand how these organs and tissues function, it is important to know the parts of whole respiration organ tracts. In fact, there are many parts of the respiration organ such as an upper and a lower organ tract including the airways, the lungs and linked blood vessels, and the muscles that enable breathing. The lower respiratory tract or lower airway is derived from the developing foregut and consists of the trachea, the bronchi and bronchioles, and lungs including alveoli [2]. Meanwhile, nose, sinuses, pharynx, or larynx are considered as the upper respiratory tracts.

In the past, there have been numerous respiratory systems developed to visualize 3D representation of internal organs. However, most of these systems do not provide deep modeling and simulation of internal organs. In this paper, we propose a hierarchical structure for representing 3D respiration organ models. Specifically, we propose a visualizing method that facilitates a representation of the 3D model organ using menu level categories. The main objectives of this study are: (1) to give names for critical regions in respiratory internal organ and (2) to construct a hierarchical tree respiratory organ based patches or regions of upper and lower respiratory organs.

The rest of the paper is organized as follows. Section 2 describes related work. Section 3 describes the proposed method in detail. Section 4 presents implementation details. Section 4 summarizes the paper and highlights future work.

2 Related Work

The systems for visualization internal organ must include all necessary components for representing the 3D respiration organs, including both upper and lower organ regions based on category level structuring menus. However, while specific systems only include an anatomical knowledge, most systems focus only on the modeling and visualizing the specific parts of the human organ such as the modeling of lungs, trachea and bronchial tree, or the specific parts of human upper or lower airway. In this section, we briefly describe these methods.

Nithiarasu et al. [3] proposed a method to explain the problems and importance of the anatomy of airflow through a human's upper airway. The authors showed the distinct sections of the upper airway based on anatomical, physiological, and pathological consideration that are divided into six segments: the nasal cavities, nasopharynx, oropharynx, laryngopharynx, larynx, and trachea. Furthermore, Kaye et al. [4] introduced an integrated approach from the prospective of functional anatomy. Specifically, the authors visualized lung functionalities using anatomy and physiology of the human body. In similar research, K. Horsfiled et al. [5] presented the techniques of bronchial categorization to measure flow in the larger airways of the human bronchial tree. The authors also gave the details about concepts related to the branching system. However, these methods are too complicated to understand and do not provide convenient visualizing methodologies.

There are several web-based respiratory systems, such as ZYGOTE BODY [6] and INNERBODY [7]. Moreover, a few other websites introduced the visualization system for the human internal organ system with menu options in order to give names for critical regions in respiratory internal organ. ZYGOTE BODY only provides the names of the whole parts of 3D human body visualization. Meanwhile, INNERBODY gives both name and category menu structure for the 2D organ models and only name modeling for 3D models. Even though, there is a considerable advancement in visualization of internal organs, most systems developed in the past do not provide the deep simulation and visualization features for 3D human internal organ models. Moreover, qualitative simulations have many drawbacks. Thus, we have introduced the visualization system for representing 3D respiratory organ model by using level category method to provide tree structure to model each respiratory internal organ's part.

3 Hierarchical Structure for Representing 3D Respiratory Internal Organ

The respiratory organ models play an important role in the proposed system. Most of these models relate to the airway system in the human body that enables us to breath including the inhaling and exhaling air in the body, the absorption of oxygen from the air in order to produce energy, and the discharge of carbon dioxide which is the byproduct of the process [8]. In order to understand the need for a specific airway simulation, it is necessary to understand the anatomy object models and anatomy of airway processes. Figure 1 shows a diagram of the whole respiratory tract of the human beings. Here, body organs involved in respiration are divided into the upper and lower respiration tracts.

The upper tract consists of main organs such as nasal, paranasal, pharynx, and larynx, while the lower tract is composed of trachea, right and left bronchi, and lungs [9]. Nasal (nasal cavity) refers to the two sides of the nose which conditions the air to be received by the other areas of respiration tract, while paranasal refers to a group of four paired air-filled spaces that surround the nasal cavity including frontal sinus, sphenoid sinus, ethmoid sinus, and maxillary sinus. Moreover, the part of digestive system is called pharynx (nasopharynx, oropharynx, and laryngopharynx), and larynx has associated cartilage to produce sound, sometimes also called a voice box. Meanwhile, trachea is the largest tube in the respiratory tract; bronchi are divided into two right and left sides that provide a passage of airway that conduct air into lungs, and the largest organs in the lower respiration track are lungs [9, 10].

In the proposed method, the whole model of the human respiratory internal organ system is collected for making the visualization system. This model includes two main parts within the internal organs, such as upper respiratory tract (e.g., paranasal sinuses, nose or nostrils, pharynx, larynx, and sometimes consists of tra-

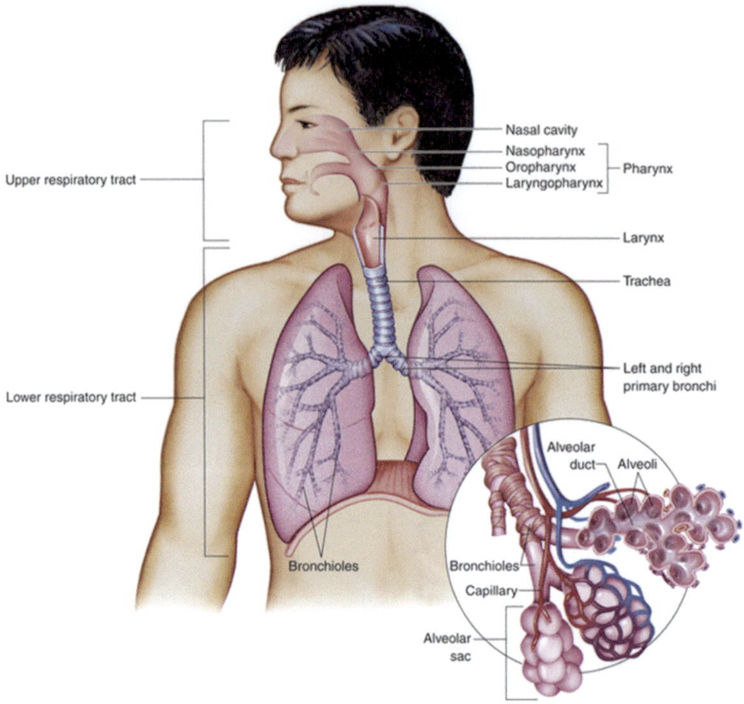

Fig. 1 Organs of the respiratory system, with a detailed view of the alveolar sac [9]

chea organ), and lower respiratory tract (e.g., trachea, bronchial tree, right lung, left lung, diaphragm, and alveoli) as shown in Fig. 2.

We developed the proposed method for the respiratory organ of the human beings by using category level structuring method, which is the branching structure that can be found throughout human anatomy and is considered as one of the most challenging structures to learn in medical education. Here, a long structure typically contains multiple structures within an organ or tissue. Understanding the relationship between individual branches and adjacent structures is important because it has a direct effect on normal and abnormal functions in a particular tissue [11]. With this proposed method in the system development, there are three main steps. First, 3D respiration organ models are partitioned into sub-parts or sub-organs by using 3D creation tool for the system visualization, and then the real color or image textures are applied. Second, we give names for critical regions, which is the way to visualize in the respiratory internal organ. Menu categories' structure and level are created in the last step for providing hierarchical option structures in visualizing processes.

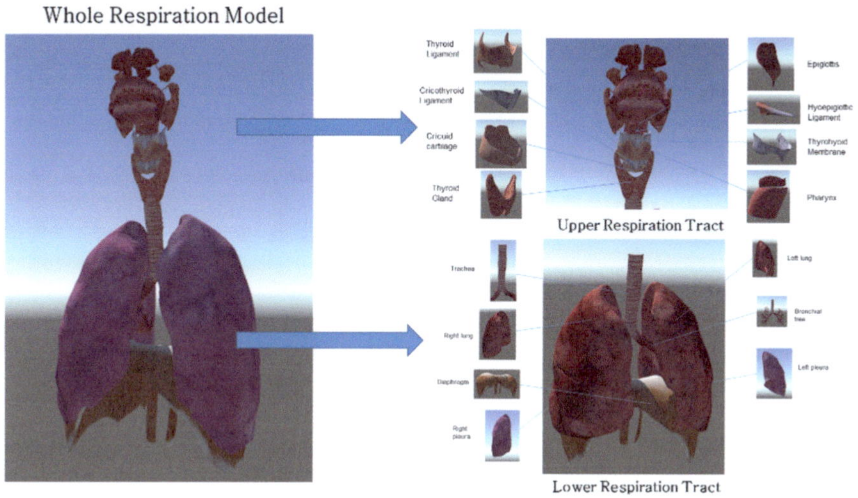

Fig. 2 The major passages and structures of the upper respiratory tract and lower respiratory tract

3.1 3D Human Respiratory Organ Partition Technique with the Real Texture

The anatomical internal organ models are of importance in the proposed system. Thus, we took important concepts from state-of-the-art books [12] as reference when partitioning the internal organs and providing names. A whole internal organ consists of many organ parts, therefore, it is divided into sub-parts based on its belonging names. To make partitions of various organ models and apply those organs with the real textures, 3D creation tool which is called Blender 3D Tool is used [13]. There are a few methods to split or subdivide the model parts including knife tool method and Boolean method. In this paper, we first use the knife tool method to partition organ models and then apply materials and textures. This method is created for the purpose of making mesh object event for connecting with Unity3D tool more easily (Fig. 3).

3.2 3D Internal Organ Name Giving

There are many types of the human internal organ system that contain organs such as cardiovascular system, digestive system, endocrine system, nervous system, reproductive system, and respiratory system. However, the respiratory system is one of the most important parts that we mentioned in our visualization system. Thus, we provide the critical names to those internal organs of the respiratory system. The reason for this is to have a better understanding about each respiratory tract's names

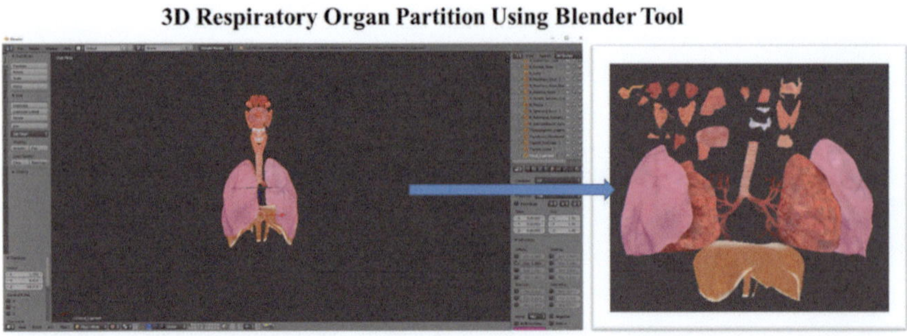

Fig. 3 Subdividing or partitioning of 3D respiratory organ model by using knife tool method in Blender 3D tool and applying those subdividing parts with the real color or image textures

for both outside and inside parts. While assigning the detailed names for all internal organ models, we followed a close reference to books and websites including "Trunk, Viscera, Lower Limb" from Atlas of Human Anatomy [12], "the Respiratory System—Structure and Function" from Bloomfield Science Museum Jerusalem website [8], and any other resources. For instance, larynx organ consists of six cartilages: three single of epiglottic, thyroid, and cricoid and three paired of arytenoid, corniculate, and cuneiform [14]. Moreover, there are still other parts of that organ as shown in Fig. 4.

In Fig. 5, we give an example of the internal organ's names containing in the whole internal organ of both upper and lower respiratory tracts such as right/left frontal sinus, nasopharynx, right/left maxillary sinus, right/left arytenoid cartilage, pharynx, epiglottis, thyroid cartilage, thyroid gland which consist of upper part, and bronchial tree, right/left lung, right left pleura, and diaphragm. We have defined all label names of the 3D respiratory human organ by using Unity3D engine tool with C# programming language. This enables to provide information of the object models when the user starts clicking.

3.3 Category Level Structuring Menu Object Method

We prepared structure method involving 3D category level in the respiratory menu system. This system has all the categories including main to the subdivided categories by using the TreeView structure. A tree structure refers to a way of representing the hierarchical files and directory structures. In addition, a concept of tree structure generally composes with the "root" at the top and the "leaves" at the bottom. Thus, as Figs. 6, 7, 8, and 9, the humanoid internal model organs are defined into the respiration tree from root to leaves such as paranasal sinuses, nasal (nose), pharynx, larynx, right/left lung, and diaphragm, etc.

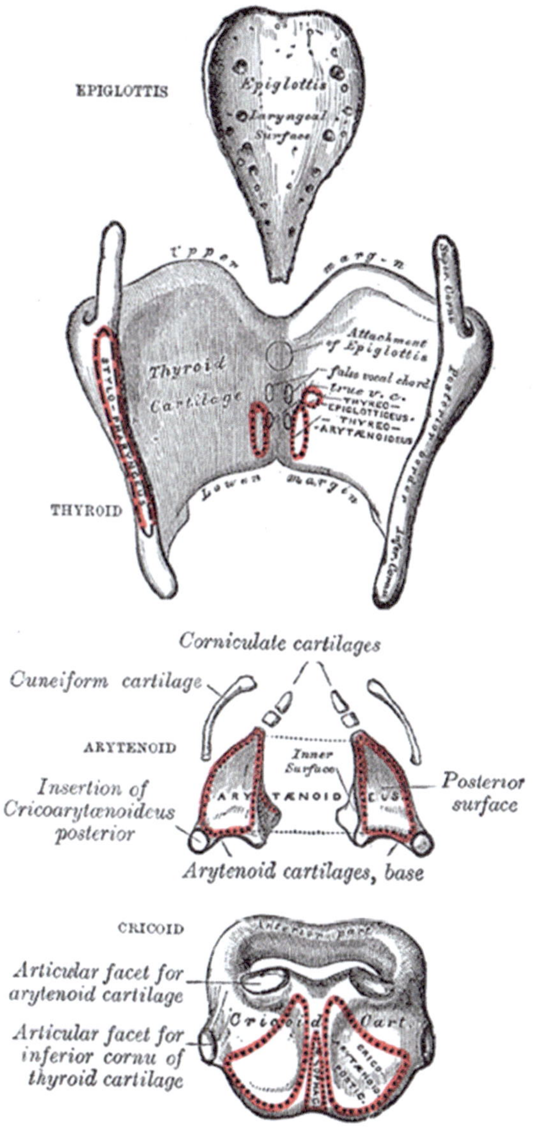

Fig. 4 Posterior view of the larynx [13]

We use the tree structure algorithm for placing and locating files (called records or keys) in a database. While the algorithm finds data by repeatedly making choices at decision points called nodes, a node consists of two branches called children. In a tree, the locations that are used to store the records are called leaves. The code syntax tree is constructed as the parsed code of source code. The base class of the syntax tree is the solution node. Every item in the solution is added to the tree structure in the

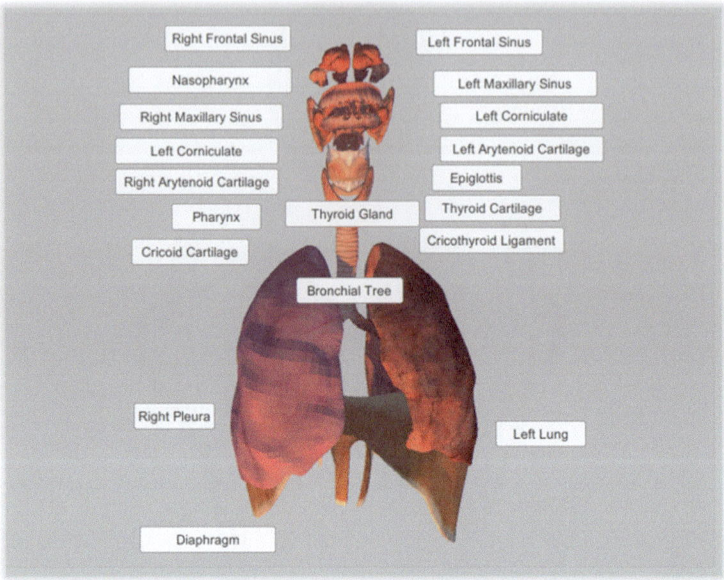

Fig. 5 It shows names for regions in the respiratory internal organ model

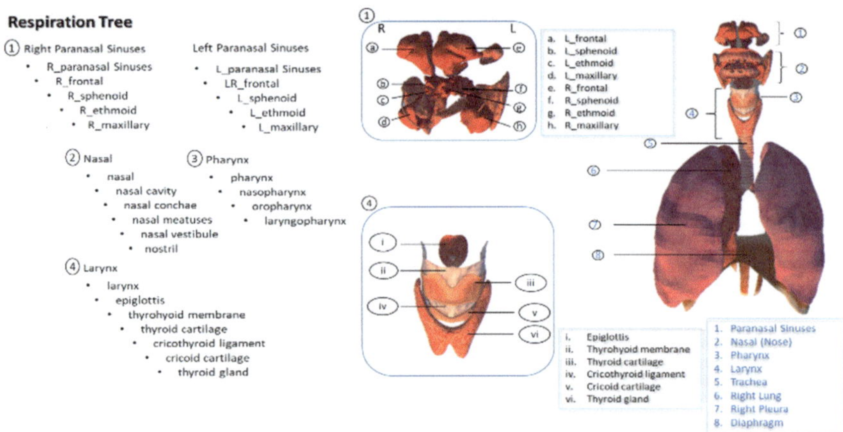

Fig. 6 The respiration tree of paranasal sinuses, nasal, pharynx, and larynx

type of hierarchical ordered system including a number of children and each child has a function such as methods of classes [17]. In computer science, a hierarchical tree structure is widely used with a set of linked nodes. By defining the section point as the node of a tree, we can get a clear context between the section points or branching structures. For example, one of the E-Learning platform courses has used tree structure method to define the obvious hierarchical characteristics. Each point

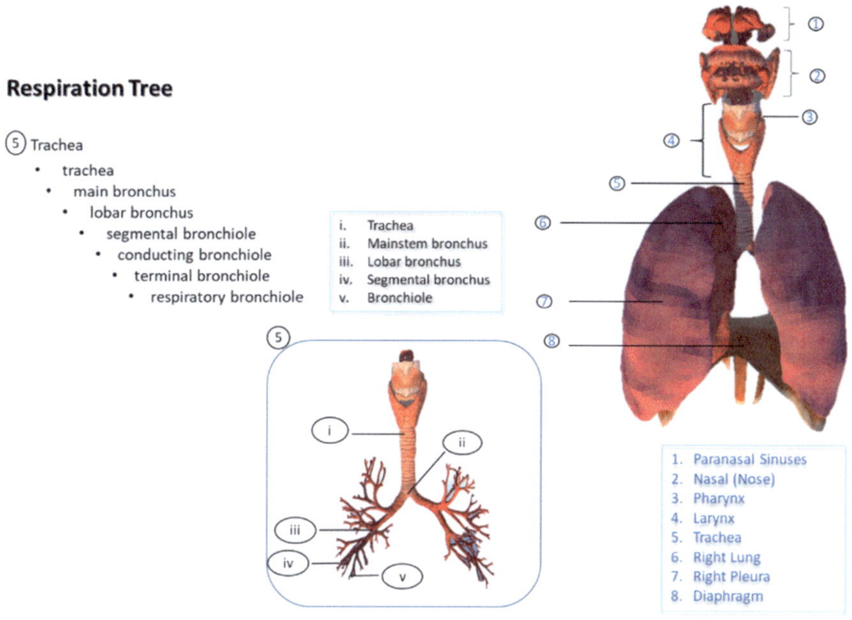

Fig. 7 The respiration tree of trachea

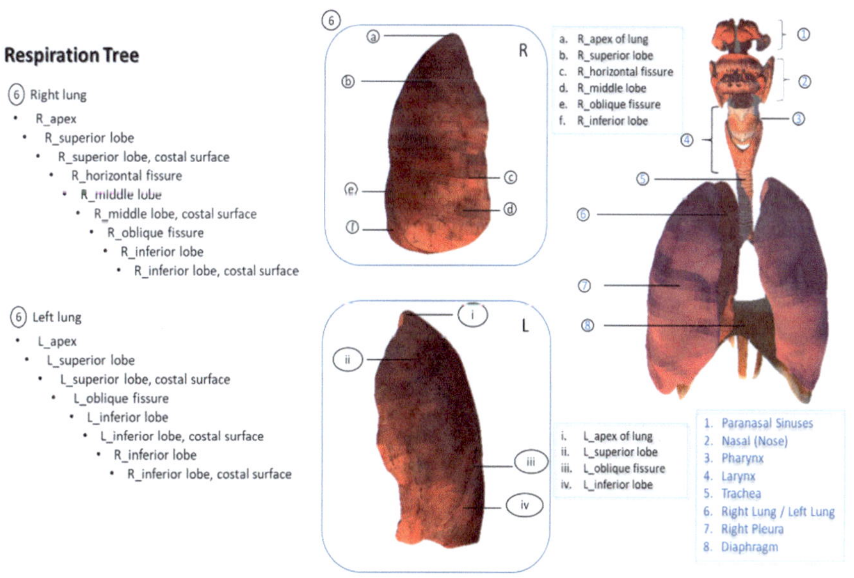

Fig. 8 The respiration tree of right and left lungs

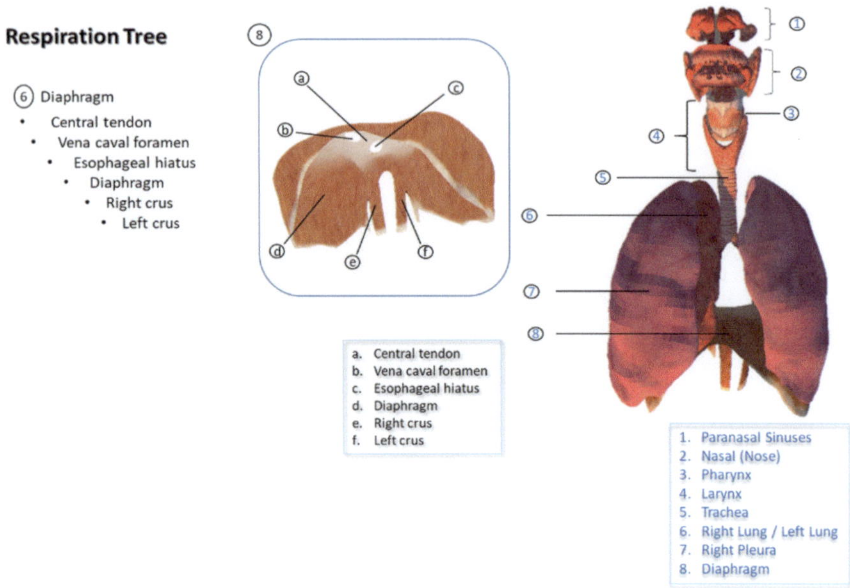

Fig. 9 The respiration tree of diaphragm

of knowledge (chapter or section) is a node of the tree, each node of the tree has the same properties, and it should be a record in the table [15].

Furthermore, we implemented some features of the TreeView category such as data binding, drag and drop, multi-select and range selection, auto scroll, and item removal to illustrate how they allow the rapid pruning of the tree when attributes are available at the node level. Data binding is a main feature in the system since it allows the user to bind the data item properties to the user interface (UI) elements in the Unity3D tool in order to display the menu items. This method gets data item properties by using index from Unity UI elements' position. Thus, it will list all properties that contain in that editor's hierarchy of UI elements by index we set. In addition, with drag and drop features, it describes the action of the selecting an object or section of text to move it (drag), and then place it (drop). Another feature is multi-selection and rang selection feature that synchronizes the option selection with Unity3D editor's hierarchy. The user can make a single or multiple select on the menu option structures with the multi-select and range selection features. Particularly, this system will provide auto scroll while expanding and the user can delete the unsatisfied menu from the hierarchy category menu list in the system (Figs. 10 and 11).

There are two ways to visualize the internal organ model: (1) To integrate between category tree structuring menu and the internal organ models, the TreeView structure in C# programming language with Unity 3D tool is used. We have created TreeViewDemo script to define the features of the structure which included dragging and dropping, selecting, and removing. With the provided selecting feature,

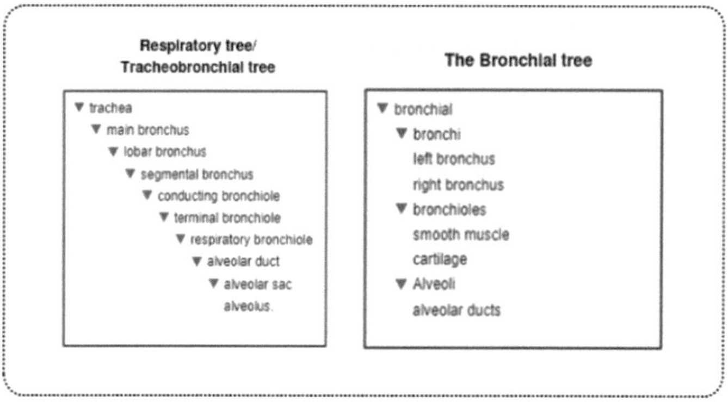

Fig. 10 (left) Tracheobronchial tree (left) describes the branching airways structure consists of trachea, bronchi, and bronchioles. And, the bronchial tree (right) refers to the names for its resemblance to the branches of a tree which include three main organs: bronchi, bronchioles, and alveoli [16]

it called method in MenuEventListener class which is responsible for integrating MeshRenderer and GameObject class of the 3D object model. We have compared names of the object menu. Then, the color and name are set, thus, it will view on object model. The process can be performed based on scripts which are added in the Unity3D tool. (2) Each part on the organ models can be clicked directly for visualizing color and names of the models (Fig. 12).

4 Implementation

We created a 3D human respiratory organ visualization system with the right category level structures that allow the user to easily change the visualization for all parts of the respiratory internal organ tracts. This system supports with 1280*720 of screen resolution. Figure 13 shows the category level structuring of 3D human respiratory organ system with level tree organ structure from parent node (root node) to children node (leave nodes) of larynx, bronchiole, lungs, trachea, etc. Thus, the users can click on the menu category or just click directly to the mesh object parts of the model on the scene to give the visualization of those internal organs' names.

5 Conclusion and Future Work

We have presented a method and detailed results of visualization system for 3D representing internal organ models which are developed for constructing a hierarchical tree respiratory organ based on patches or regions of the upper and lower respiratory

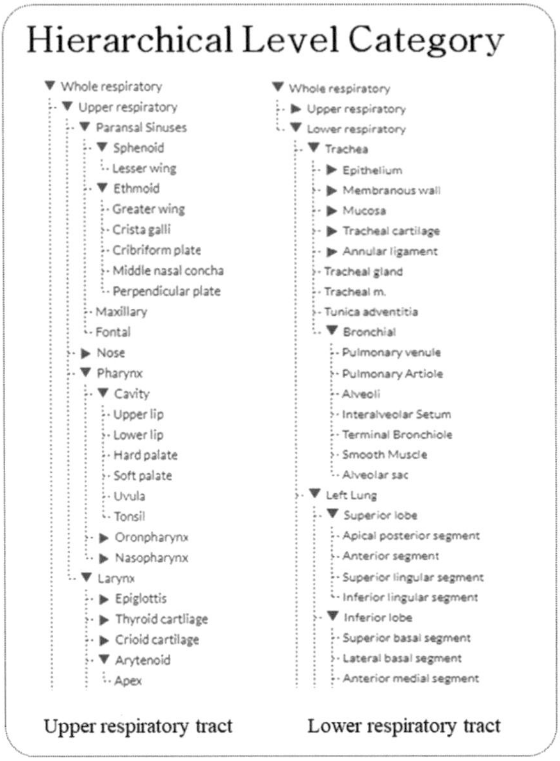

Fig. 11 (right) Category level tree structuring architecture of 3D human internal organ models based on upper and lower respiratory tracts. This structure is divided from top to bottom, means from root to leave node by using Unity3D tool with C# programming language

organs, and giving names for critical regions in the 3D respiratory internal organ. Moreover, even though this system provides useful features to the user, it still needs to be improved and updated to operate more effectively and efficiently in computer science sector, especially, medical science field. Thereby, for our future work, we will define more functional properties for each critical region in the respiratory organs. Particularly, animation of the respiratory internal organ will be added into the next updated system. In addition, symptoms and diseases belonging to those organ regions will be added. This system will be updated to provide more convenient, and deep visualization for the users.

Acknowledgments This research was supported by the Ministry of Science and ICT (MSIT, Korea), under the Information Technology Research Center (ITRC) () support program (IITP-2017-2015-0-00448) supervised by the Institute for Information & Communications Technology Promotion (IITP) () and by the Technology Innovation Program (or Industrial Strategic Technology Development Program) (10065694, International Standardization of 3D Digital Human Modeling) funded by the Ministry of Trade, Industry & Energy (MOTIE, Korea).

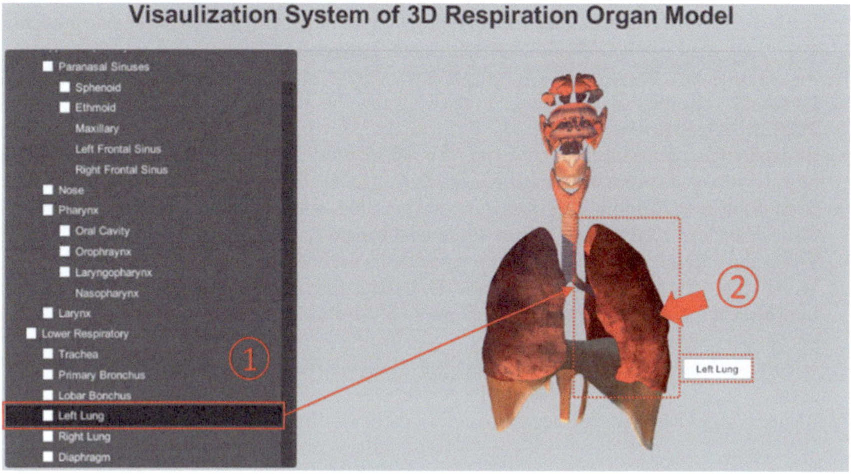

Fig. 12 (1) The category tree structure of the menu event occurs to visualize the models and (2) direct object event occurs to visualize the organ models

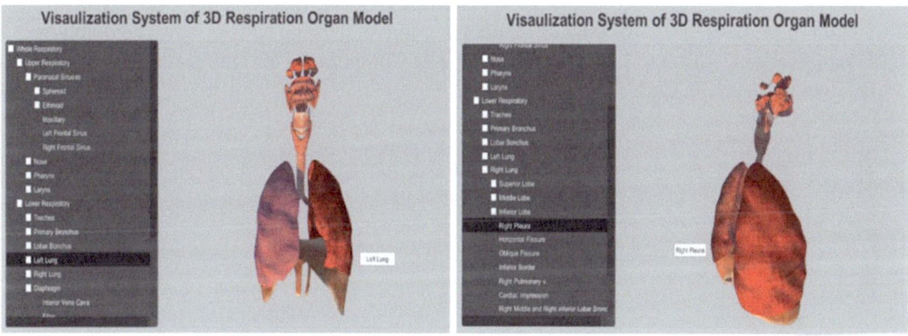

Fig. 13 The results of a visualization system for representing 3D respiratory organ models by using category level structuring method of TreeView

References

1. A. R. Lambert, P. O'Shaughnessy, M. H. Tawhai, E. A. Hoffman, C. L. Lin, "Regional deposition of particles in an image-based airway model: large-eddy simulation and left-right lung ventilation asymmetry", the PubMed Central (PMC), vol. 45, No. 1, pp. 11–25, 2011
2. "Drawing in and Processing Air: How the structures of the Lower Respiratory System Work", Visible Body, https://www.visiblebody.com/learn/respiratory/lower-respiratory-system, 2007–2017
3. P. Nithiarasu, O. Hassan, K. Morgan, N. P. Weatherill, C.Fielder, H. Whittet, P. Ebden, K. R. Lewis, "Steady flow through a realistic human upper airway geometry", International Journal for Numerical Methods in Fluids, vol. 57, issue 5, pp. 631–651, 2008
4. J. Kaye, D. Metaxa, J. R. Clarke, B. Webber, "Lung Modeling: Integrating Anatomy and Physiology", the First International Symposium on Medical Robotics and Computer-Assisted Surgery (MRCAS-94), 1994

5. K. Horsfield, G. Dart, D. E. Olson, G. F. Filley, G. Cumming, "Models of the human bronchial tree", Journal of Applied Physiology, vol. 31, no. 2, August 1971
6. ZYGOTE BODY, https://www.zygotebody.com, Zygote Media Group, Inc., 2014–2017
7. Innerbody, "Respiratory System", http://www.innerbody.com/anatomy/respiratory, HOW-TOME- DIA, INC. 1999–2017
8. Bloomfield Science Museum Jerusalem, "The Respiratory System – Structure and Function", https://www.mada.org.il/en/about/engineer/challenge/respiratory-system, 2013
9. A. M. Ramirez, "Respiratory System", https://musculoskeletalkey.com/respiratory-system-3/, Musculoskeletal Key| Fastest Muscu- losketletal Insight Engine, 2016
10. Wikipedia, "Respiratory tract", https://en.wikipedia.org/wiki/Respiratory_tract, 2017
11. W. A. Mattingly, J. H. Chariker, R. Paris, D. Chang, J. R. Pani, "3D Modeling of Branching Structures for Anatomical Instruction", Journal of Visual Languages and Computing, vol. 29, pp. 54–62, August 2015
12. R. Putz, R. Pabst, "Trunk, Viscera, Lower Limb", Atlas of Human Anatomy, vol. 2, pp. 090–097, 2008
13. Blender, http://pennstatehershey.adam.com/content.aspx?productId=112&pid=28&gid=0–00141
14. Wikipedia, "Larynx", https://en.wikipedia.org/wiki/Larynx, 2017
15. Z. Chen, J. Shi, Z. Li, J. Chen, "An E-learning Platform Design Based on Tree Structure", International Conference on E-Health Networking, Digital Ecosystems and Technologies, pp. 237–240, 2010
16. "Bronchial Tree", Innvista, http://innvista.com/health/anatomy/bronchial-tree/, 2016
17. S. Sarala, M. Deepika, "Unifying clone analysis and refactoring activity advancement towards C# applications", 4th International Conference on Computing, Communications and Networking Technologies (ICCCNT), July. 4–6, 2013

An Effective Method for Detecting Outlying Regions in a 2-Dimensional Array

Ki Yong Lee and Young-Kyoon Suh

Abstract As sensing devices and simulation programs are widely used, a large amount of output data is being generated in the form of a 2-dimensional (2D) array. To facilitate the post data processing, it is of critical importance to find anomalous or outlying elements in that array with little human intervention. In this paper, we propose an effective method for locating outlying regions in a 2D array, in which a group of adjacent elements in its entirety deviate significantly from the entire array. To find such outlying regions, we divide the array into small subarrays and build a regression model for each subarray. We then cluster subarrays that are adjacent to each other and have similar regression models into larger subarrays. After the clustering, we detect relatively small clusters as outlying regions. Our experiments confirm the effectiveness of the proposed method.

Keywords Outlier detection · Collective outliers · 2-dimensional array analysis

1 Introduction

With the advance and prevalence of scientific measurement and sensing devices, a large amount of data is being generated in the form of a 2-dimensional (2D) array. Also, many simulation programs output their simulation results as 2D arrays. One of the important tasks for analysis of such array data is to find anomalous or outlying elements in an array. Finding outlying elements is very useful to identify measurement errors, singular observations, or diverged simulations. In an array, an outlying

K. Y. Lee
Division of Computer Science, Sookmyung Women's University,
Seoul 04310, Republic of Korea
e-mail: kiyonglee@sookmyung.ac.kr

Y.-K. Suh (✉)
School of Computer Science and Engineering, Kyungpook National University,
Daegu 41566, Republic of Korea
e-mail: yksuh@knu.ac.kr

W. Lee and C. K. Leung (eds.), *Big Data Applications and Services 2017*,
Advances in Intelligent Systems and Computing 770,
https://doi.org/10.1007/978-981-13-0695-2_5

element indicates one whose value is significantly different from its neighbor elements. Thus, we need to consider both the position and the value of an element in that array to determine its outlierness.

There has been a rich body of literature to find outliers given a set of data objects. Especially, spatial outliers are those whose values considerably differ from their spatial neighbors. Several approaches in the body were proposed to find such a spatial outlier. The first is to find data objects that are very far from their neighbors as compared to the other objects [1]. However, this approach is not applicable to array data considered in the paper. The second is to find data objects whose values are notably divergent from their neighbors [2]. In this approach, most works have a similar assumption that normal objects have values that are uniform or homogeneous. In other words, they assume that the values of normal objects are drawn from the same distribution. Accordingly, they determine the outlierness of a data object based on how far its value is from the average of its neighbors. However, in many real-world applications, normal values themselves often internally have some pattern of change. Hence, the average-based outlier judgment may lead to incorrect results.

In this paper, we propose an effective method for detecting outlying regions in a 2D array such that its elements internally reveal some pattern of change. Note that rather than picking up an outlying element(s), our focus is on finding outlying "*regions*" in which a group of adjacent elements in its entirety deviate significantly from the entire array.

2 Proposed Outlying Region Detection Method

Let A be a 2-dimensional $n \times m$ array, where n and m are the numbers of rows and columns, respectively. $A[i, j]$ is the element in A located at the ith row and jth column. A region in an array A is a group of adjacent elements in A. An outlying region in A is a region in A whose elements have a significantly different pattern from its neighboring regions.

The proposed method is largely divided into five steps: (1) *Array partitioning*: we divide the input array A into equal-sized square subarrays with side k. (2) *Regression model construction*: for each square subarray, we construct a regression model that represents the pattern of change of the values in the subarray. (3) *Model tagging*: we convert each regression model into a vector using its coefficients and tag the vector to the corresponding subarray. (4) *Subarray clustering*: we cluster into larger subarrays the subarrays that are adjacent to each other and have similar regression models. (5) *Outlying region determination*: we identify the clusters with very few elements and report these small clusters as outlying regions. Each of the five steps is elaborated below.

(1) **Array partitioning**: Given the input array A, we divide A into a number of small subarrays, each of which is a k-sized square region, where k is a user-specified number. The smaller k, the finer-grained the outlying regions are found.

(2) **Regression model construction**: To model the pattern of change of values in each subarray, we build a regression model for each subarray. In this paper, we use the multiple linear regression model [3], which is represented as follows. Here, a_0, a_1, a_2, and a_3 are the coefficients of the model, X_1 and X_2 are independent variables that represent the row and column index, respectively, and Y is a dependent variable that represents the predicted value of $A[X_1, X_2]$.

$$Y = a_0 + a_1X_1 + a_2X_2 + a_3X_1X_2 \tag{1}$$

(3) **Model tagging**: After we obtain a regression model $Y = a_0 + a_1X_1 + a_2X_2 + a_3X_1X_2$ for a square subarray with size k, we tag the vector (a_0, a_1, a_2, a_3) to the corresponding subarray. This vector is called *model vector*. In this paper, we measure the similarity between two regression models using the Euclidean distance between their model vectors. In other words, for the two regression models $Y = a_0 + a_1X_1 + a_2X_2 + a_3X_1X_2$ and $Y = b_0 + b_1X_1 + b_2X_2 + b_3X_1X_2$, their distance is computed as $((a_0 - b_0)^2 + (a_0 - b_0)^2 + (a_0 - b_0)^2 + (a_0 - b_0)^2)^{1/2}$.

(4) **Subarray clustering**: We cluster subarrays that are adjacent and close to each other in terms of their model vectors. The purpose of this clustering is to combine into a larger, single subarray the adjacent subarrays that show a very similar or the same pattern of change. In this paper, we use an agglomerative clustering method [4] to cluster subarrays. When merging two subarrays, we select two subarrays that are adjacent to each other and have the smallest distance between their model vectors. This clustering is continued until there are no more two clusters whose distance between their model vectors is below a specified threshold, T_s.

(5) **Outlying region determination**: Finally, we determine outlying regions using the subarray clusters obtained in Step (4). For each cluster, we measure the size of the cluster as the number of elements in that cluster. If the size of a cluster is less than T_s, we report that small cluster as an outlying region. In other words, we output a group of adjacent elements as a single outlying region if its size is small and its pattern of change is different from the neighboring regions. If we consider outlying regions with size less than 1/10 of the entire size of the array, we may set T_s to $nm/10$.

The proposed algorithm can find outlying regions in a 2D array that show a different pattern of change from their neighboring regions and of small sizes, from the specified Steps (1)–(5). Note that an outlying region cannot be large by its definition.

3 Experiments

We now evaluate the effectiveness of the proposed method. We implemented the proposed method using R. For the datasets used in the experiments, we created synthetic 2D arrays that internally had some pattern of change among their elements.

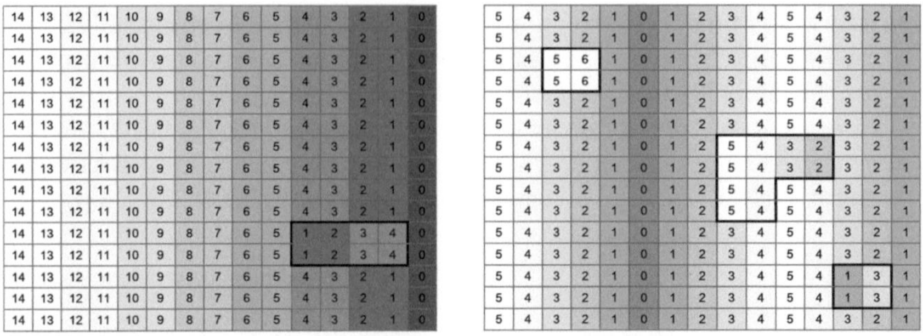

(a) An 2D array with a monotonically - decreasing pattern

(b) An 2D array with a repeatedly increasing or decreasing pattern

Fig. 1 Two synthetic 2D array data used in the experiments

Figure 1a shows the 2D array data for the first experiment to evaluate the correctness of the proposed method. The size of the array is 15×15, and we have set $k = 2$. In Fig. 1a, the elements in that array show a decreasing pattern as they move to the left. Note that the smaller the value, the darker the element is. Obviously, the rectangle region at rows 11–12 (from the top) and columns 11–14 (from the leftmost) is an outlying region, because its pattern of change strongly deviates from that of its neighboring regions. Indeed, that region cannot be detected by existing methods that determine the outlierness of a region based on the difference between its average and its neighboring average. That is because the average of the region does not differ much from that of its neighbors. The proposed method, on the contrary, can correctly find the outlying region, due to the distinguished regression model of the region.

Figure 1b shows another 2D array data used in the second experiment, where the elements are repeatedly increasing or decreasing while moving to the left. In this case, we can easily identify three outlying regions with different sizes and shapes, which are marked with bold borders. Similar to the first experiment, the proposed method can detect all the three outlying regions correctly, while existing average-based methods fail to detect them.

4 Conclusion

In this paper, we have proposed an effective method for detecting outlying regions in a 2D array. In order to determine the outlierness of a region, instead of comparing its average and its neighbors' average, the proposed method compares its regression model and its neighbors' regression models. Through the experiments, we have shown that the proposed method can detect outlying regions more effectively than existing average-based methods.

Acknowledgements This research was supported by the National Research Foundation of Korea (NRF) grant funded by the Korea government (MSIP) (No. 2015R1C1A1A02037071).

References

1. Chawla, S., Sun, P.: Slom: a new measure for local spatial outliers, Know. Inf. Syst. 9(4), 412–429 (2006)
2. Telang, A., Deepak, P., Joshi, S., Deshpande, P., Rajendran, R.: Detecting localized homogeneous anomalies over spatio-temporal data, Data Min. Knowl. Disc. 28(5–6), 1480–1502 (2014)
3. Cohen, J., Cohen, P., West, S. G., Aiken, L. S.: Applied multiple regression/correlation analysis for the behavioral sciences, 3rd ed., Lawrence Erlbaum Associates (2003)
4. Fahad, A., et al.: A Survey of Clustering Algorithms for Big Data: Taxonomy and Empirical Analysis, IEEE Transactions on Emerging Topics in Computing, 2(3), 267–279 (2014)

An Effective Recall-Oriented Information Retrieval System Evaluation

Justin JongSu Song, Jiyoung Lim, Chris Soo-Hyun Eom and Wookey Lee

Abstract The objective recall-oriented retrieval tasks such as patent retrieval, legal search, and e-Discovery is to find all relevant documents. These information retrieval tasks are usually conducted by the domain experts so the retrieval cost might be high. However, the existing evaluation metrics for recall-oriented retrieval tasks have obvious limitations which consider only high recall value. If the relevant documents are at the lower position in the total ranked result, the user should check the whole result up to the end. Therefore, the evaluation measure for recall-oriented retrieval tasks is strongly required reducing the review efforts. In this paper, we study the feature of the various evaluation metrics according to the ranks of the query result in the recall-oriented retrieval tasks.

Keywords Recall-oriented retrieval · Evaluation measure · Patent retrieval

1 Introduction

Intellectual property including patent and trademark has been caught general attention more than before, since the impact and the role of the businesses have become crucial nowadays [8, 9]. Thus the importance to find related patents and prior arts is bigger and bigger in the companies and governments. However the patent search and the prior art search have revealed the characteristics far different from those of the conventional searches, since it places just importance on retrieving each and every

J. J. Song · J. Lim · C. S.-H. Eom · W. Lee (✉)
Department of Industrial Engineering, Inha University, Incheon, South Korea
e-mail: trinity@inha.edu

J. J. Song
e-mail: jaegal83@inha.edu

J. Lim
e-mail: ljy@inha.edu

C. S.-H. Eom
e-mail: sunching@inha.edu

W. Lee and C. K. Leung (eds.), *Big Data Applications and Services 2017*,
Advances in Intelligent Systems and Computing 770,
https://doi.org/10.1007/978-981-13-0695-2_6

item all, which can be achieved by total recall [5–7]. However, the patent search also needs to be incorporated from the keyword and graph search in database community [1, 4, 12] and probabilistic optimization results [3, 10]. There have been classified the typical four patent search types based on risks involved in patent search [2]. At first, State-of-the-Art determines business decision about whether or not invest capital and develop technology in question by conducting this search in that technology field. This search starts in the very beginning or before investment so there is relatively low risk involved. Secondly, patentability places their goal to protect companys technology by being patentable. If this search is conducted poorly, the company might be at risk such as competitor inflow or changing development plan. Thirdly, Freedom-to-operate should be conducted thoroughly otherwise a technology which company has already produced and released might infringe another companys technology. The cost of patent litigation is enormous. It is the closest step to recent patent litigation matters in the world. Lastly, Validity is the most comprehensive type of patent search that involves corporate decision making and includes risks that can lead to other legal issues. Table 1 shows that types of patent search according to risks and attributes of information retrieval system.

As we have seen above, it is important to consider risks originated by patent retrieval. Therefore, we define two concepts of risk arising from the inherent characteristics of patent search and from the information retrieval (IR) system perspective. Recall-oriented retrieval tasks such as patent retrieval, legal search and e-Discovery place great importance on finding all relevant documents [6, 9, 11]. These information retrieval tasks are usually conducted by the domain experts so the retrieval cost might be high. However, the existing evaluation metrics have obvious limitations for measuring recall-oriented retrieval systems because these metrics consider only high recall value. The retrieval cost and the relevance should be considered in the case of the same recall threshold. Let us consider an example as follows. Each and every search system has to find 3 relevant documents out of 10 targets and there are four search engines called System 1 through 4. Assuming that System 1 gets the first and the second results {1, 2} out of 10 relevant documents, which means the third

Table 1 Types of patent search based on risks and search result quality

	Low	<———————————>		High
Risk	State-of-the-Art	Patentability	Freedom-to-Operate	Validity
Recall	Patentability	Validity	State-of-the-Art	Freedom-to-Operate
Precision	State-of-the-Art	Patentability	Freedom-to-Operate	Validity
$\|N\|$	Validity	Freedom-to-Operate	Patentability	State-of-the-Art
$\|n\|$	Patentability	Validity	Freedom-to-Operate	State-of-the-Art

has not been retrieved. System 2 has resulted {4, 5, 6} rankings, System 3 {1, 5, 9}, and finally System 4 gets {1, 2, 9}, respectively. Which are summarized in Table 1 in more detail. In this example, which one gives the best result? Even if the System 1 finds the relevant 2 results as rank 1 and 2, but lost one. The other 3 systems find all 3 targets but different rankings. Disregarding system 1, all the 3 systems can be evaluated differently with respect to each and every thresholds. Notice that system 2 will be finalized the search effort earlier and thus it should be regarded highly, since it is very much natural that the number of the target documents will be increased up to millions or trillions as the real patents and prior arts. Up to now, there have been seldom tackles on this issue. Assuming that the result at any rate will be retieved but appeared in the last ranked position, which will torture the reviewer to finalize the search effort. Therefore the recall should be satisfied as well as the last relevant item should also be listed as earlier as possible. In order to achieve the previous goals together, there needs an evaluation measure to be devised so that the relevance as well as the search effort together can be fairly evaluated. By doing that, the search result evaluation measure for the high recall related issue that should also reduce the users retrieval effort. In this paper, we study the features of the system according to the ranking of the query results in the recall-oriented retrieval environment.

2 Information Retrieval Evaluation Metrics

With the explosion of information, it is important for those information to be managed and handled effectively. In this situation, it is also essential to test how well IR system satisfies information needs for users. It requires the more appropriate standards based on specific IR systems objectives and users information needs. A variety of IR evaluation metrics have been researched and explored for how closely correlated they are with user satisfaction criteria, how meaningful average values are, and how much power they have to discriminate among retrieval results.To evaluate the performance of recall-oriented search system, we should not consider the recall only. If we only consider recall rate, the number of retrieved results can be maximized for improving recall. That is, the number of total documents which reviewer should check increase and it should take a great deal of time and cost. We may see the various metrics for measuring information retrieval systems performance and we also explain how to calculate each evaluation metrics.

2.1 Mean Average Precision (MAP)

When we try to compare the quality of IR systems for each single queries, the most standard among the TREC community is MAP [11]. It is assumed that all documents based on users query and search goal should be in a result set.

$$MAP = \frac{\sum_{q=1}^{Q} AP(q)}{Q} \tag{1}$$

where, Q is the number of queries (q) and AP is the average precision.

2.2 nDCG

The information retrieval is an interactive process and a user also expects to put in less effort to retrieve information. For example, a user wants to find the most relevant document among results or a ranked position in the result. However, precision and recall can only address binary relevance or non-relevance. In this perspective, nDCG is designed for situations in which non-binary relevance. Jrvelin and Kekinen [2] introduced the Discounted Cumulative Gain(DCG) for measuring effectiveness of ranking quality. This metric has been used as the standard evaluation measure when graded relevance values are available. There are two assumptions in using Discounted Cumulative Gain.

- Highly relevant documents appear earlier in a search result than marginally relevant documents.
- Relevant documents ranked lower position are less useful for the user, since it is less likely to be examined.

nDCG is to calculate the cumulative gain values. Cumulative gain is accumulated from the relevance of the top ranked document to the last document. Cumulative Gain (CG) is defined as:

$$CG[k] = \sum_{r=1}^{k} G[i] \tag{2}$$

where $G[i]$ is the graded relevance of the document of ith ranked.

The next step of $nDCG$ is to penalize documents appearing lower in a result set. The graded relevance is typically reduced by $log_2(1+k)$. So, the discounted cumulated gain is defined as following:

$$DCG[k] = \sum_{r=1}^{k} \frac{G[i]}{log_2(1+i)} \tag{3}$$

To compare search systems performances, the DCG should be normalized based on users search satisfaction. That is, the ideal ordering which maximizes CG should be considered.

The normalized discounted cumulated gain ($nDCG$), which is a final step, is finally defined as:

$$nDCG = \frac{DCG[k]}{IDCG[k]} \tag{4}$$

2.3 Normalized Recall

Normalized recall is one of the representative recalloriented IR evaluation measures. This metric is proposed for measuring the effectiveness in ranking documents relative to the best and worst ranking cases. The best case is that all relevant documents are retrieved at the top of the list and the worst case is that all relevant documents are retrieved at the bottom of the list. In the worst case, the user should check full collection for gaining all relevant documents. Normalized recall is one of the recall oriented IR evaluation metric. It is defined as:

$$R_{norm} = 1 - \frac{\sum r_i - \sum i}{n(N-n)} \tag{5}$$

where r_i is the rank at which the ith relevant document is retrieved, N is the size of the total documents retrieved, and n is the number of relevant documents.

2.4 PRES (Patent Retrieval Evaluation Score)

It is important to retrieve all relevant document such as patent retrieval or legal retrieval. For those recall-oriented contexts, PRES (patent retrieval evaluation score) has been recently processed [7]. It is based on the same idea as the normalized recall but PRES is different from the worst case. Unlike R_{norm}, user could find the relevant document just after the total result list. It is calculated as follows:

$$R_{norm} = 1 - \frac{\frac{\sum r_i}{n} - \frac{n+1}{2}}{N_{max}} \tag{6}$$

where r_i is the rank at which the ith relevant document is retrieved, N is the size of the total documents retrieved, n is the number of relevant documents, and R is Recall (number of relevant retrieved docs in the 1st N_{max} docs).

This metric considers either promoting a system that retrieves as many documents as possible or privileging a system that retrieves relevance documents early.

3 Evaluating Recall-Oriented Retrieval Systems

To evaluate the performance of recall-oriented search system, we should not consider the recall only. If we only consider recall rate, the number of retrieved results can be maximized for improving recall. That is, the number of total documents which reviewer should check increase and it should take a great deal of time and cost.

We analyze the limitation of the existing evaluation metrics and we will consider why we need a new evaluation metric for recall-oriented retrieval tasks. Table 2 shows the search results of each system (system 1 to 4). The total number of search results(N) is 10 and each system have different ranking results. If relevant documents are retrieved according to following orders $\{4, 5, 6\}$ (system 2) or $\{1, 5, 9\}$ (system 3), which system would be preferred by users? A reviewer prefer system 2 to system 3. Because the reviewer can obtain all relevant document at the sixth document and there is no need to see more (importance of review effort). In addition, in case of system 3 and system 4, the reviewer can find all relevant documents with same search time but system 4 might be preferred because it retrieves the relevant documents a little faster than system 3 (importance of average precision). In case of system 1 and system 3, system 3 might be preferred since system 1 fail to achieve total recall (importance of recall). While considering high recall, the existing evaluation metrics such as MAP, Recall, F, nDCG, and PRES cannot reflect all of these conditions. MAP is not sufficient to be used in total recall case because it gives high score to a system which can retrieve relevant documents at the higher rank. In case of recall, it fails to distinguish between system 2, 3 and 4 which system is more effective. In case of F-measure, even though this metric considers both precision and recall, it also fails to distinguish system 2, 3 and 4. Because F-measure has same concept of recall. nDCG is not suitable for recall-based search because the purpose of nDCG is to find highly relevant documents earlier not focusing on finding all relevant documents. Finally, PRES [7], which is representative recall-oriented evaluation metric, is insufficient to meet all of these conditions. It also fails to distinguish between system 2 and system 3. Moreover, it evaluated system 4 better than system 2. The reason is that PRES cannot reflect users review effort since it only calculates distance between best and actual retrieval cases.

Table 2 Performance of different evaluation values with different IR systems

	Ranks of rel. docs	MAP	Recall	nDCG	F'_4	PRES
Sys. 1	{1, 2}	0.667	0.667	1	0.586	0.744
Sys. 2	{4, 5, 6}	0.383	1.0	0.501	0.879	0.7
Sys. 3	{1, 5, 9}	0.578	1.0	0.664	0.879	0.7
Sys. 4	{1, 2, 9}	0.778	1.0	0.88	0.879	0.37

4 Conclusion

In this paper, we scrutinize a new evaluation metric for recall-oriented application. The existing evaluation measurement methods for recall-oriented retrieval tasks have obvious limitations which consider only high recall value not the effort to reach the required result. Thus the existing evaluation metrics have failed to distinguish each system performance but a proper evaluation measure should consider not only high recall but also review efforts.

References

1. Anne, C., Mishra, A., Hoque, M.T., Tu, S.: Multiclass patent document classification. Artificial Intelligence Research **7**(1), 1 (2017)
2. Järvelin, K., Kekäläinen, J.: Ir evaluation methods for retrieving highly relevant documents. In: Proceedings of the 23rd annual international ACM SIGIR conference on Research and development in information retrieval. pp. 41–48. ACM (2000)
3. Kim, J., Lee, W., Song, J.J., Lee, S.B.: Optimized combinatorial clustering for stochastic processes. Cluster Computing **20**(2), 1135–1148 (2017)
4. Lee, R.C., Cuzzocrea, A., Lee, W., Leung, C.K.: An innovative majority voting mechanism in interactive social network clustering. In: Proceedings of the 7th International Conference on Web Intelligence, Mining and Semantics. p. 14. ACM (2017)
5. Li, C., Wang, Y., Resnick, P., Mei, Q.: Req-rec: High recall retrieval with query pooling and interactive classification. In: Proceedings of the 37th international ACM SIGIR conference on Research & development in information retrieval. pp. 163–172. ACM (2014)
6. Lim, J.: Total recall evaluation metric: Trem for patent retrieval case. Master thesis, Inha Univ. (2017)
7. Magdy, W., Jones, G.J.: Pres: a score metric for evaluating recall-oriented information retrieval applications. In: Proceedings of the 33rd international ACM SIGIR conference on Research and development in information retrieval. pp. 611–618. ACM (2010)
8. Samuelson, P.: Supreme court on design patent damages in samsung v. apple. Communications of the ACM **60**(3), 26–28 (2017)
9. Venugopal, A.V.: Inadvertent disclosures of inventions in social media affecting patent rights. Computers & Security **72**, 136–144 (2018)
10. Yuan, Y., Lian, X., Chen, L., Yu, J.X., Wang, G., Sun, Y.: Keyword search over distributed graphs with compressed signature. IEEE Transactions on Knowledge and Data Engineering **29**(6), 1212–1225 (2017)
11. Zhang, H., Lin, W., Wang, Y., Clarke, C.L., Smucker, M.D.: Waterlooclarke: Trec 2015 total recall track. In: TREC (2015)
12. Zhao, X., Liang, J., Dang, C.: Clustering ensemble selection for categorical data based on internal validity indices. Pattern Recognition **69**, 150–168 (2017)

Chentry: Automated Evaluation of Students' Learning Progress for Entry Education Software

Jeong-Hun Kim, Jong-Hyeok Choi, Uygun Shadikhodjaev, Aziz Nasridinov and Ki-Sang Song

Abstract In this paper, we introduce Chentry, which is an automated evaluation system for Entry education software to assist programming education. We base our system on Entry platform which is a block based programming platform aimed at teaching coding for learners with easy-to-grasp visual concepts. Our system uses logs of Entry to evaluate student performance and by use of Levenshtein distance provides assessment of student performance.

Keywords Entry education software · Automated evaluation · Code analysis

1 Introduction

Traditionally, education has been an important part of any country's development agenda since it is the cornerstone of the country's long-term development. With the innovations happening in information technologies and computer science in particular, education has been affected by these new technologies. With the advent of Internet and web technologies, more and more spheres of our life have been going online with most of the services having dual representation: in the reality and virtual reality. Education is also a service in this respect and has been constantly affected by

J.-H. Kim · J.-H. Choi · U. Shadikhodjaev · A. Nasridinov
Department of Computer Science, Chungbuk National University, Cheongju, Korea
e-mail: etyanue@chungbuk.ac.kr

J.-H. Choi
e-mail: leopard@chungbuk.ac.kr

U. Shadikhodjaev
e-mail: uygun@chungbuk.ac.kr

A. Nasridinov
e-mail: aziz@chungbuk.ac.kr

K.-S. Song (✉)
Department of Computer Education, Korea National University of Education, Cheongju, Korea
e-mail: kssong@knue.ac.kr

W. Lee and C. K. Leung (eds.), *Big Data Applications and Services 2017*,
Advances in Intelligent Systems and Computing 770,
https://doi.org/10.1007/978-981-13-0695-2_7

the advent of the web technologies. Developments in the sphere of education have resulted in the evolvement of new platforms and universities, foreign and domestic alike, have been introducing elements of web based education into their curriculums with some of these universities running totally online programs that have eliminated classroom as a physical medium needed to gather students and instructors into one place.

In this paper, we discussed a proposed system named Chentry which is a web based system developed for educational purposes for learners of online learning platform Entry (www.playentry.org) that uses block based coding to teach learners easy programming concepts through visual techniques. The platform is free and offers log based analysis for its users thereby enabling its usage for the teachers and students at the same time. Our proposed system named Chentry by combining the words "checking" and "entry" uses these mentioned advantages of Entry system and provides convenient views for students and teachers to create assignments and monitor student progress. Features of the Chentry system are proposed and discussed with concrete illustrations and visualizations. Our system consists of several modules and views with use of Levenshtein distance to evaluate student performance.

2 Related Study

Traditional educational setting with teacher–student interaction happening in class has been replaced or complemented by E-learning and learning management systems (LMS), Intelligent tutoring systems (ITS), and Adaptive educational hypermedia system (AEHS) to name a few [1]. These new technologies have led to accumulation of data and were applied to various stages of education such as student behavior modeling, student performance modeling, assessment of students' work and assignments, student modeling which also includes student cognitive state analysis, student support and feedback generation during study process, curriculum design, domain knowledge representation, curriculum sequencing, and other tasks that are prevalent in the educational domain [2]. These developments have raised interest from both the practitioners and experts in both educational and computer science domains, which has resulted in multitude of systems and platforms built for the newly explored domain of computer based education systems that are aimed at automation of the education process.

Automatic grading of student performance in terms of assignments and other data is one part of the efforts made to increase the effectiveness of educational systems. In [3], authors have devised a program that automatically evaluates students' programming assignments that was used in the Rutgers University to assess the work of students routinely for some 300–600 assignments weekly. The system was tested and used extensively which resulted in over 20,000 student programming assignments being graded with an error rate of only 5%. In another research [4], authors have suggested using student answers for classifying their performance against fuzzy rule models that use both expert knowledge and knowledge extracted from data. Later research has focused on using different technologies of machine learning like hierarchical clustering algorithm [5] to evaluate the performance of students in an Information Technology (IT) course and has concluded overall predictability of applied techniques for evaluation of student performance. These kinds of automation techniques can lead to effective cost benefits with increased number of enrollments and allow instructors focus on other aspects of teaching process while having useful insights from the mentioned approaches. This kind of relationship was inspected by Wilcox [6], which has concluded that careful design of student performance evaluation system can lead to student performance improvement and at the same time reduce the amount of resources that are dedicated to such task thereby bringing efficiency into the educational process.

3 Chentry

In this section, we describe Chentry in detail. We first provide an overview of the proposed system in Sect. 3.1, and then in the subsequent subsections, we explain each module of the proposed system in detail.

3.1 Overview

Figure 1 shows overall structure of the proposed Chentry system. It contains three modules: view module, main module, and database module. The view module displays the result of automated evaluation in both teacher's and students' web pages. The main module is responsible to check, make structure, and evaluate the code. In database module, all details of the teachers and students along with their progress information are stored. More details of each module will be provided in the subsequent subsections.

3.2 View Module

The main purpose of the view module is to provide an easy-to-use environment for both teachers and students who use Chentry system. Later, implementation details of the view environment will be given in Chap. 4. Figure 2 shows details of the view module.

As we can observe from the figure, Chentry can be used by one teacher or group of teachers. At initial stage, a teacher provides the correct answer code for a coding task that was given to a class of the students. Further, students, in turn, provide a solution for the coding task. Chentry analyzes the code provided by both parties and generates an evaluation report. In the view module, teachers and students are provided with

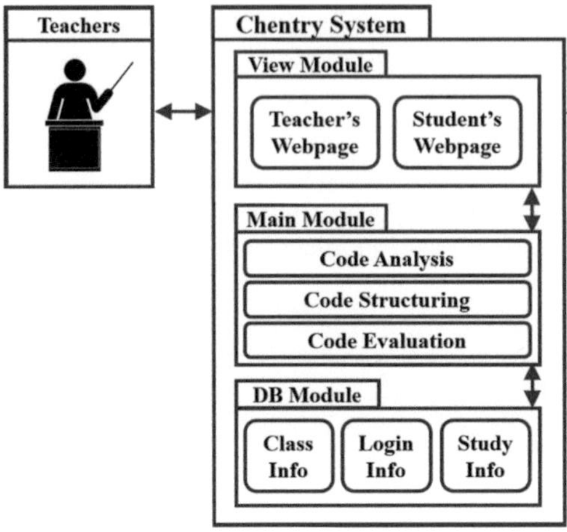

Fig. 1 Overall structure of Chentry

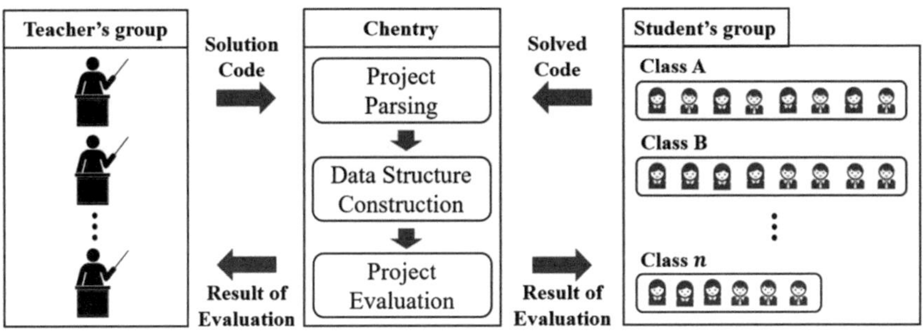

Fig. 2 Detailed representation of the view module

personal web pages, where they can learn about the evaluation report. The details of the code analysis and evaluation report will be given in the next subsection.

3.3 Main Module

The main module has three components that are needed to generate an evaluation report. These components are code analysis, code structuring, and code evaluation. This subsection will provide details of the key components of the main module.

Once the codes are received from both parties (i.e., correct answer code and solved code), the main module first performs a code analysis procedure. From Fig. 3a, we can see that Entry scene (i.e., project) contains objects and block code. Here, the block code defines a code procedure that defines objects' behavior. For example, in Fig. 3a, the code procedure indicates an action, where an object moves 10 times 10

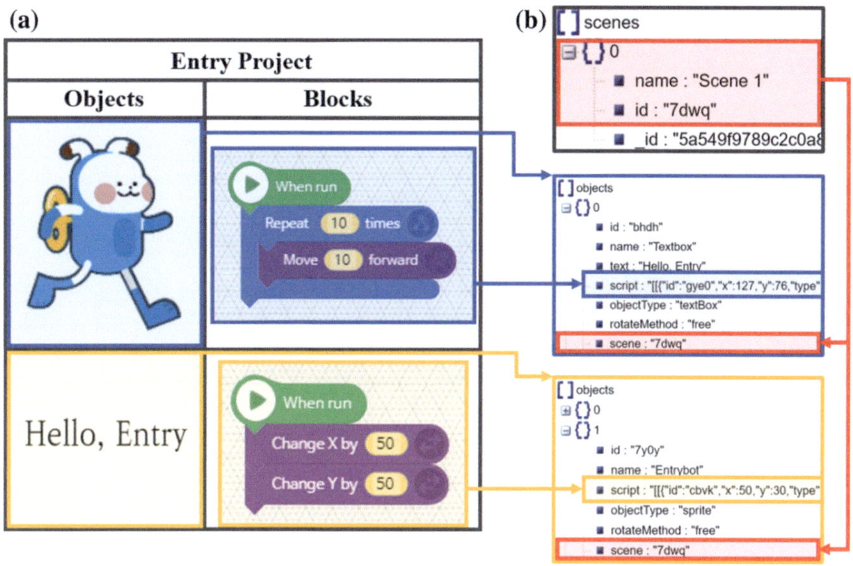

Fig. 3 Extraction of object and block code from entry scene

units forward. Figure 3b shows a JavaScript Object Notation (JSON) representation of the Entry project. By observing the JSON representation, we recognize that every Entry scene has a name, defined by a user, and a unique identification code. This unique identification code is included in all objects of the scene, which enables us to extract objects corresponding to the specific Entry scene. The JSON representation also provides us with the object related information, such as unique identification code, name, and other details, and the block code related information, such as unique identification code, type of code, parameters, units and their values.

Figure 3 provides evidence that JSON code has patterns. In other words, all JSON codes corresponding to the specific Entry scene have the same structure. By using these characteristics, we propose a code structure representation procedure. Specifically, we propose to convert JSON representation of each Entry scene to a tree structure, called Chentry tree. Figure 4a shows Entry scene and 4b shows its Chentry tree representation. From the tree representation, we can observe that the root node is always a *project node*. Further, *object nodes* represent objects of the scene. *Block group nodes* characterize the block codes of the scene, and *block nodes* denote detailed procedure of the corresponding block codes. Finally, other inner details of the block codes are represented in an *inner block node*. The main feature of Chentry tree is that it can efficiently represent the whole Entry scene in a tree structure that can be later used for the analysis task.

Once Chentry tree is constructed for both parties (i.e., correct answer code and solved code), we can perform code analysis by comparing their code and generate evaluation. In order to perform code analysis, we use Levenshtein distance measure-

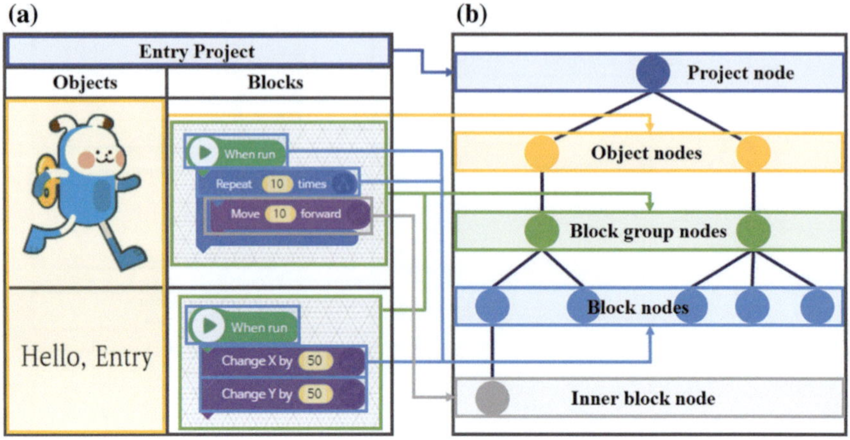

Fig. 4 Converting JSON code to a tree structure

ment. Originally, it represents the minimum number of operations (i.e., insertion, deletion, substitution) required to make given two strings A and B equal. In the proposed method, we use this characteristic of Levenshtein distance and calculate the minimum number of modifications of the required blocks to be the same as the correct answer code. An example of application of Levenshtein distance to Chentry is shown in Fig. 5. Figure 5a demonstrates correct answer code provided by a teacher who performs basic calculation procedure. Figure 5b depicts solved code provided by a student. In the student's solved code, we can notice that several parts of the code do not match with correct answer code provided by a teacher. By analyzing the code of the both parties using Levenshtein distance, we can learn that there are minimum four operations that must be performed in order to correct the student's code.

In the proposed system, Levenshtein distance enables us to detect minimum number of operations needed to correct the code. In order to convert this into an expressive evaluation report, we propose to calculate student progress through the following formulas.

$$Individual\,block\,group\,match\,(\%) = 100 * \left(1 - \frac{Levenshtein\,block\,distance}{MAX\,(No.of\,theacher's\,block,\,No.of\,student's\,block)}\right) \tag{1}$$

$$Overall\,block\,group\,match\,(\%) = 100 * \left(1 - \frac{Levenshtein\,block\,distance}{\sum MAX\,(No.of\,theacher's\,block,\,No.of\,student's\,block)}\right) \tag{2}$$

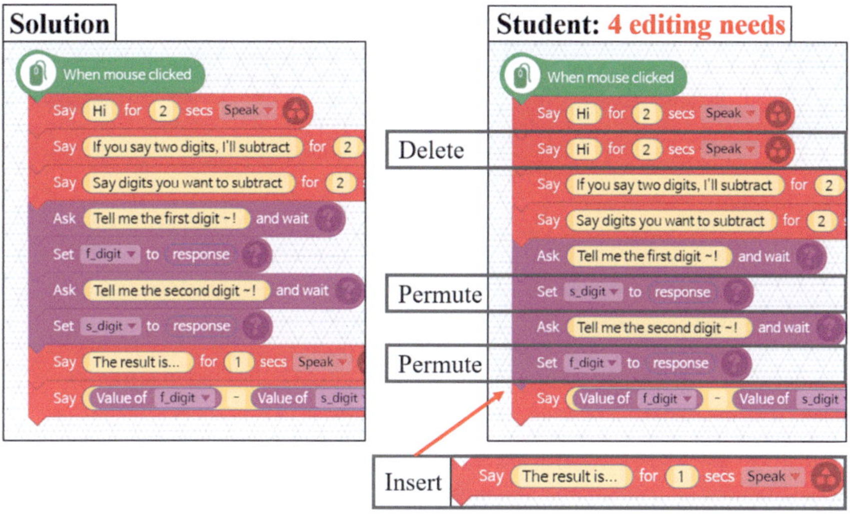

Fig. 5 Application of Levenshtein distance to Chentry

Here, if the block type is the same but only the block parameters are incorrect, then we perform the calibration 0.5 times rather than once or evaluate block type.

3.4 DB Module

The last module of the Chentry is a database (DB) module. The DB module stores and manages information related to the teacher and students. Specifically, it stores information related to the login credentials of the teacher as well as class information and progress of each student. As for students' information, the DB module manages their login credentials, progress, and study path. Storing information related to the teachers and students enables us to perform knowledge discovery procedure. In other words, the information collected during the years of Chentry usage may become potentially huge, and thus, if we apply machine learning or data mining techniques, we can obtain valuable insights useful for both teachers and students. For example, by analyzing the learning path of the large number of students, we can recommend a correct path to the students who experience the same problem over and over again.

4 Implementation

In this chapter, we explain implementation details and user interface (UI) of the proposed Chentry system. The implementation details of the proposed system are

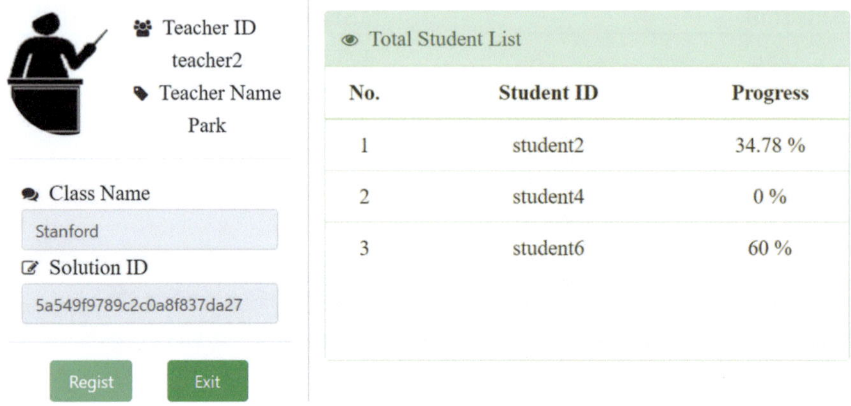

Fig. 6 Teachers' view in Chentry system

as follows. The web interface of the proposed Chentry system is developed using JavaServer Pages (JSP). The code analysis of teachers' and students' codes with Levenshtein distance is performed using Java programming language. In addition, we used MySQL database for storing teacher and student information.

For the sake of simplicity, we divide the explanation of UI into two parts, such as teachers' view and students' view. Further details of each view will be given in the subsequent subsections.

4.1 Teachers' View

Teachers' view provides a convenient access to evaluate students' progress for the teachers. Figure 6 demonstrates the details of UI for teachers' view. On the left side of the window, we can see the teacher's details, including his unique identification code, name, and name of class he is teaching. Recall from Chap. 3 that each teacher provides the task and correct answer code to the Chentry that is later used to check the students' progress. Teacher can register the task using UI provided by Chentry. Once this procedure is performed, the teacher receives the solution identification code for the task. On the right side of the window, we can see a list of students along with their progress in percentage calculated using formulas (1) and (2).

4.2 Students' View

Students' view offers a convenient view for each student to check his or her progress compared to the teachers' progress. Figure 7 shows UI for students' view in Chentry.

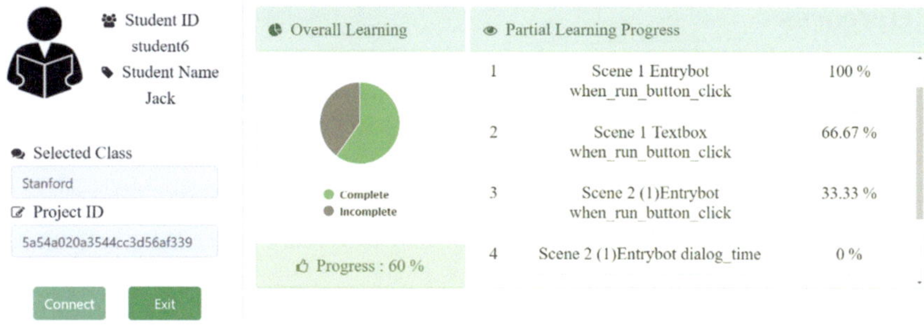

Fig. 7 Students' view in Chentry

On the left side, we can see personal details of the student, such as unique identification code, name, and name of the class he has joined. Each student can look through various tasks provided by the teacher and select one s/he is interested in to solve. Once the student selects the task, Chentry connects him with the task and assigns solution code. After solving the code provided by the teacher, s/he must save the code in Entry system that automatically triggers a generation of progress report. Recall from Chap. 3 that progress report is generated by comparing the JSON code of teacher and that of the student using Levenshtein distance. Once the report is generated, the student can easily check by clicking on the report. On the right side of Fig. 7, we can see the details of the progress report generated by Chentry. One of the main features of Chentry is that it can produce a detailed report that points out to the exact flaws in students' code as shown in Fig. 7 in percentage.

5 Conclusion

In this paper, we have introduced Chentry, which is an automated evaluation system for Entry education software. Chentry automatically generates reports regarding students' coding progress based on teacher and student code comparisons. For code comparisons, we used Levenshtein distance that can measure the distance in JSON codes. The advantage of the proposed system is that teachers and students can easily use it in order to check the learning progress. Recall from Chap. 3 that, the information collected during the years of Chentry usage may become potentially huge, and thus, if we apply machine learning or data mining techniques, we can obtain valuable insights useful for both teachers and students. Thus, in the future, we develop Chentry that can adapt big data.

Acknowledgements This work was supported by the National Research Foundation of Korea (NRF) grant funded by the Korea government (Ministry of Science, ICT & Future Planning) (No. 2017R1A2B1012909).

References

1. Romero, C., & Ventura, S. (2010). Educational data mining: a review of the state of the art. IEEE Transactions on Systems, Man, and Cybernetics, Part C (Applications and Reviews), 40(6), 601–618.
2. Peña-Ayala, A. (2014). Educational data mining: A survey and a data mining-based analysis of recent works. Expert systems with applications, 41(4), 1432–1462.
3. Morris, D. S. (2003, November). Automatic grading of student's programming assignments: an interactive process and suite of programs. In Frontiers in Education, 2003. FIE 2003 33rd Annual (Vol. 3, pp. S3F-1). IEEE.
4. Rasmani, K. A., & Shen, Q. (2005, May). Subsethood-based fuzzy rule models and their application to student performance classification. In Fuzzy Systems, 2005. FUZZ'05. The 14th IEEE International Conference on (pp. 755–760). IEEE.
5. Rana, S., & Garg, R. (2016, February). Application of Hierarchical Clustering Algorithm to Evaluate Students Performance of an Institute. In Computational Intelligence & Communication Technology (CICT), 2016 Second International Conference on (pp. 692–697). IEEE.
6. Wilcox, C. (2015, February). The role of automation in undergraduate computer science education. In Proceedings of the 46th ACM Technical Symposium on Computer Science Education (pp. 90–95). ACM.

Constrained Big Data Mining in an Edge Computing Environment

Carson K. Leung, Deyu Deng, Calvin S. H. Hoi and Wookey Lee

Abstract High volumes of wide varieties of valuable data of different veracity can be easily generated or collected at a high velocity from various big data applications and services. A rich source of these big data is the Internet of Things (IoT), which can be viewed as a network of sensors, mobile devices, wearable devices, and other "things" that are capable to operate within the existing Internet infrastructure. As a popular data science task, frequent pattern mining aims to discover implicit, previously unknown and potentially useful information and valuable knowledge—in terms of sets of frequently co-occurring items—embedded in these big data. Existing frequent pattern mining algorithms mostly run serially on a single local computer or in distributed and parallel environments on computer clusters, grids, or clouds. Many of these algorithms return large numbers of frequent patterns, of which only some may be of interest to the user. In this paper, we present a constrained big data mining algorithm that (i) focuses the mining to those frequent patterns that are interested to the users and (ii) runs in an edge computing environment, in which computation is performed at edges of the computing network.

Keywords Big data · Big data applications and services · Data analytics · Data mining · Data science · Edge computing · Frequent patterns · Knowledge discovery in databases

1 Introduction

With advances in technology, high volumes of a wide variety of data of different veracity (e.g., uncertain data [23, 27]) can be easily generated or collected at a high velocity from rich sources of data—such as devices from the Internet of Things

C. K. Leung (✉) · D. Deng · C. S. H. Hoi
University of Manitoba, Winnipeg, MB, Canada
e-mail: kleung@cs.umanitoba.ca

W. Lee
Inha University, Incheon, South Korea

W. Lee and C. K. Leung (eds.), *Big Data Applications and Services 2017*,
Advances in Intelligent Systems and Computing 770,
https://doi.org/10.1007/978-981-13-0695-2_8

(IoT) [6, 31] like sensors, actuators, mobile devices (e.g., mobile phones), wearable devices (e.g., smart watches), and other "things" that are capable to operate within the existing Internet infrastructure—in various real-life big data applications and services [19] such as bioinformatics [15], games [5, 7], social media [3, 8], social networks [17, 22, 24, 25, 28], smart worlds [18], Web [26], and wireless sensor systems [9]. Embedded in these big data are useful information and valuable knowledge.In other words, these big data demonstrated the characteristics that are commonly known as 5V's of big data:

1. Volume, which focuses on the quantity of big data (e.g., in terms of petabytes (PB) = 10^{15} bytes, Exabyte (EB) = 10^{15} bytes, zettabyte (ZB) = 10^{21} bytes, yottabyte (YB) = 10^{24} bytes);
2. Variety, which focuses on the differences in types, contents or formats of big data (e.g., bioinformatics data like DNA sequences, games data like series of moves, social media data like audio or videos posted on social networking sites like Facebook or Twitter, Web data like Web logs, wireless sensor data like measurements collected by sensors);
3. Veracity, which focuses on the quality of big data (e.g., uncertainty, messiness, or trustworthiness of big data);
4. Velocity, which focuses on the speed at which data are collected or generated (e.g., around 6000 tweets are tweeted on Twitter per second); and
5. Value, which focuses on the usefulness of big data (e.g., "big data is the new oil").

Hence, *data science* [11]—which applies data mining, database, machine learning, visualization, mathematical, and/or statistical techniques to big data for *data analytics* [20]—are in demand.

As a popular data science task, *frequent pattern mining* [14, 16] aims to discover sets of frequently co-occurring items. The mined patterns help data scientists understand consumer behavior, know the web surfing patterns, and get an insight about popular events. Early frequent pattern mining algorithms (e.g., Apriori [1]) mostly execute the mining process in serial on a single local computer. To improve algorithmic efficiency, later algorithms perform the computation in distributed and parallel environments by using computer clusters or grids. As volumes of big data can go beyond the ability of commonly-used software to capture, manage, and process within a tolerable elapsed time, new forms of processing, mining and analyzing data—such as MapReduce-based frequent pattern mining algorithms [4] that usually run in clouds—have been developed to enable enhanced decision making, insight, knowledge discovery, and process optimization. In general, cloud computing makes use of a network of remote servers hosted on the Internet to store, manage, process, and mine the data. To further speed up the mining process for the big data, we present in this paper a frequent pattern mining algorithm that uses *edge computing* [29, 30], which supports IoT concepts by facilitating the computation in proximity to end-users or data sources.

While many existing algorithms discover frequent patterns, most of them provide little or no support for user focus when mining data. However, in many real-life applications and services, the users or data scientists may have some particular phenomena in mind on which to focus the mining (e.g., medical analysts may want to find only those sensor data belonging to patients suspected to suffer from asthma instead of all the patients, business analysts may want to find popular checkpoints for consumers of a certain age group instead of all their consumers). Without user focus, the data scientists often need to wait for a long period of time for numerous frequent patterns, out of which only a tiny fraction may be interesting to the data scientists. This implies long and unnecessary computation, as well as lots of data transmission (between the data sources and data centers in the cloud), even in the cloud computing environment. Hence, *constrained mining* [21]—which aims to discover those frequent patterns that satisfy the user-specified constraints—is needed. As a preview, with edge computing, our algorithm pushes the constraints to IoT devices or networking services between these end devices and data centers in clouds so as to avoid unnecessary computation and reduce data transmission.

Our *key contribution* of this paper include our algorithm for mining frequent patterns satisfying user-specified constraints from big data applications and services using IoT devices in an edge computing environment. Evaluation results show that our algorithm efficiently finds frequent patterns by pushing user-specified constraints early in the mining process and closer to the end users, and thus avoiding unnecessary computation and reducing data transmission.

The remainder of this paper is organized as follows. Next section review some related works. Section 3 presents our algorithms for mining frequent pattern from big data with IoT devices in an edge computing environment. Section 4 gives our evaluation results. Finally, conclusions are drawn in Sect. 5.

2 Related Works

Many existing frequent pattern mining algorithms were serial algorithms. For instance, Apriori-based algorithms [1] rely on a generate-and-test paradigm to discover frequent patterns from transaction datasets by first generating candidates and then checking their actual frequency (i.e., occurrences) against the dataset. To improve algorithmic efficiency, other serial frequent pattern mining algorithms (e.g., tree-based algorithms [12]) have been developed.

In addition, there are also distributed and parallel algorithms [26, 32]. They find frequent patterns by using either one of the following multi-processor systems:

- systems with multiple processors that have access to a *shared memory* with which information is exchanged among processors, or
- systems with multiple processors that have their own private memory (i.e., *distributed memory*) with which for information is exchanged via message passing among the processors.

These algorithms are mainly implemented using one of the following architectures:

- Open Multi-Processing (OpenMP), which usually applies parallelism within a multi-core node; or
- Message Passing Interface (MPI), which usually applies parallelism between multiple nodes.

These algorithms run in computer clusters or grids. *Cluster computing* usually involves a group of distributed or parallel computers that are interconnected through high-speed networks such as local area networks; *grid computing* coordinates heterogeneous networked loosely coupled computers. Each computer in the grid may perform a different task.

To handle big data, MapReduce-based frequent pattern mining algorithms [4] read input data, divide them into several partitions (sub-problems), and assign them to different processors. Each processor executes the "map" function on each partition by taking a pair of $\langle key_1, value_1\rangle$ and returning a list of $\langle key_2, value_2\rangle$-pairs as an intermediate result, which is then shuffled and sorted. In other words, the "map" function can be defined as follows:

$$\text{map: } \langle key_1, value_1\rangle \rightarrow \text{list of } \langle key_2, value_2\rangle, \tag{1}$$

where (i) key_1 and key_2 are keys in some domains, and (ii) $value_1$ and $value_2$ are their corresponding values in some domains. Then, each processor executes the "reduce" function on a single key_2 from this intermediate result by combining, aggregating, summarizing, filtering, and/or transforming the list of values associated with a given key_2 (for all keys). As a result, a single (aggregated or summarized) $value_3$ is returned. In other words, the "reduce" function can be defined as follows:

$$\text{reduce: } \langle key_2, \text{list of } value_2\rangle \rightarrow value_3, \tag{2}$$

where (i) key_2 is a key in some domains, and (ii) $value_2$ and $value_3$ are its corresponding values in some domains. Many of these algorithms are run in (public, private, or hybrid) clouds. *Cloud computing* [2] involves a group of interconnected and virtualized computers to provide on-demand services such as the following:

- Function-as-a-service (FaaS),
- Infrastructure-as-a-service (IaaS),
- Network-as-a-service (NaaS),
- Platform-as-a-service (PaaS),
- Software-as-a-service (SaaS), and
- Mining-as-a-service (MaaS) [13].

In contrast, our algorithm uses *edge computing*, which supports IoT concepts by facilitating the computation in proximity to end-users or data sources (i.e., performing the computation at edges of the computing network).

To focus the mining, *constrained* frequent pattern mining algorithms [10, 21] allow the users or data scientists to use a rich set of SQL-style constraints to specify their interest for guiding the mining process so that only those frequently occurring sets of market basket items that satisfy the user-specified constraints are found. This avoids unnecessary computation for mining those uninteresting frequent itemsets. Besides market basket items, the set of constraints can also be imposed on items, events or objects in other domains. However, many existing constrained frequent pattern mining algorithms were not designed for handling big data. In contrast, our algorithm is designed to handle big data.

3 Constrained Mining with Edge Computing

Nowadays, big data grow exponentially—especially in volume—partially due to the availability, accessibility and popularity of IoT devices. For instance, big data can be collected on merchandise items that have been purchased by consumers through their mobile phones. Mining on these big purchasing data helps reveal interesting patterns such as consumer behavior, popularity of merchandise items, and/or effectiveness of mobile ads. Similarly, big data can be collected when users checked in for a place or an event via their smart watches. Mining on these activity tracker data helps reveal trajectories of popular locations or events. Moreover, big data can also be generated by monitoring sensors or fitness trackers. Mining on these big web data (health related data) helps reveal the health conditions of users.

Recall that, when applying cloud computing to mine frequent patterns, high volumes of data are usually transmitted from IoT devices to data centers in the cloud. In the cloud, data are partitioned and distributed among several worker nodes so that sets of "map" and "reduce" functions can be applied and data mining process can be executed in parallel.

In contrast, our algorithm uses edge computing, which makes good use of the edge of the network by facilitating the computation in proximity to end-users. It stores, manages, processes, and analyzes data on either of the following locations:

- IoT devices, or
- networking services between these end devices and data centers.

To elaborate, our edge-computing constrained frequent pattern mining pushes user-specified constraints to the IoT devices, in which constraints are checked. The following are some instances:

- If data scientists are interested in some but not all users (e.g., users of a certain age or gender—say, male users of age 16 or above), then only data from those users are examined.
- If data scientists are interested in some promotional merchandise items (e.g., dairy products), then only data about only those promotional merchandise items are examined.
- If data scientists are interested in some locations within a specific region (e.g., western Canada), then only data for those locations are examined.
- If data scientists are interested in some special events (e.g., BigDAS conference), then only data for those events are examined.

Then, depending on the IoT devices, our algorithm applies the following approaches:

- If the IoT devices are capable for computing the frequency counts of relevant data (e.g., interested users, items, locations, or events), then summary information (e.g., $\{a, b, c\}$ occurs together 10 times) are transmitted from the IoT devices to their local networking services that lie between the IoT devices and the usual data centers (used in cloud computing).
- If the IoT devices are not that powerful, relevant raw transactional data are transmitted from the IoT devices to their local networking services where the frequency counting takes place.

The local networking services then mine for locally frequent patterns (i.e., patterns that are frequent with respect to the data or information from one or more IoT devices served by the local networking services). Afterwards, our algorithm takes the union of these locally frequent patterns to form global candidate patterns. By counting the frequency of these global candidates on the local networking services, transmitting the frequency counts of these candidates to other devices (or a master node) and summing the frequency counts, our algorithm discovers globally frequent patterns in the network.

4 Evaluation

As *unconstrained* algorithms first find all patterns then verify if the patterns satisfy the user-specified constraints, their runtimes were unchanged regardless of the selectivity of constraints. In contrast, our *constrained* algorithm pushes the constraints into the mining process. Hence, restrictive constraints (with a few patterns satisfying the constraints) led to short runtimes and low volumes of data or information transmission. When the constraints were relaxed, more patterns satisfied the constraints. As such, runtimes and volumes of transmitted data or information transmission increased proportionally to the selectivity of the constraints.

When compared *cloud-computing based* algorithms, our *edge-computing based* algorithm appears to be more efficient and practical because computations (i.e., big data mining and analysis) are performed closer to end-users. Hence, our algorithm reduces the latency and network bandwidth, enables geographic focus, and increases reliability and security.

Moreover, our algorithm also appears to be scalable with volumes of big data when mining frequent patterns that satisfy user-specified constraints in an edge-computing environment.

As ongoing work, we are conducting more exhaustive evaluation examining different aspects of the effectiveness of our constrained big data mining algorithm.

5 Conclusions

In this paper, we presented a constrained mining algorithm that (i) focuses the mining to those frequent patterns satisfying the user-specified constraints and (ii) runs in an edge computing environment, in which computation is performed at edges of the computing network—i.e., near the data sources on local IoT devices or through local networking services that lie between the data sources and data cloud centers.

Acknowledgements This project is partially supported by (i) Natural Sciences and Engineering Research Council of Canada (NSERC) and (ii) University of Manitoba.

References

1. Agrawal, R., Srikant, R.: Fast algorithms for mining association rules in large databases. In: VLDB 1994, pp. 487–499 (1994)
2. Bierzynski, K., Escobar, A., Eberl, M.: Cloud, fog and edge: cooperation for the future? In: FMEC 2017, pp. 62–67. IEEE (2017)
3. Braun, P., Cuzzocrea, A., Doan, L.M.V., Kim, S., Leung, C.K., Matundan, J.F.A., Singh, R.R.: Enhanced prediction of user-preferred YouTube videos based on cleaned viewing pattern history. Procedia Computer Science 112, pp. 2230–2239. Elsevier (2017)
4. Braun, P., Cuzzocrea, A., Jiang, F., Leung, C.K., Pazdor, A.G.M.: MapReduce-based complex big data analytics over uncertain and imprecise social networks. In: DaWaK 2017. LNCS, vol. 10440, pp. 130–145 (2017)
5. Braun, P., Cuzzocrea, A., Keding, T.D., Leung, C.K., Pazdor, A.G.M., Sayson, D.: Game data mining: clustering and visualization of online game data in cyber-physical worlds. Procedia Computer Science 112, pp. 2259–2268. Elsevier (2017)
6. Braun, P., Cuzzocrea, A., Leung, C.K., Pazdor, A.G.M., Tanbeer, S.K.: Mining frequent patterns from IoT devices with fog computing. In: HPCS 2017, pp. 691–698. IEEE (2017)
7. Brown, J.A., Cuzzocrea, A., Kresta, M., Kristjanson, K.D.L., Leung, C.K., Tebinka, T.W.: A machine learning system for supporting advanced knowledge discovery from chess game data. In: IEEE ICMLA 2017, pp. 649–654 (2017)
8. Choudhery, D., Leung, C.K.: Social media mining: prediction of box office revenue. In: IDEAS 2017, pp. 20–29. ACM (2017)
9. Cuzzocrea, A., Grasso, G.M., Jiang, F., Leung, C.K.: Mining uplink-downlink user association in wireless heterogeneous networks. In: IDEAL 2016. LNCS, vol. 9937, pp. 533–541 (2016)
10. Cuzzocrea, A., Leung, C.K., MacKinnon, R.K.: Mining constrained frequent itemsets from distributed uncertain data. Future Generation Computer Systems 37, pp. 117–126. Elsevier (2014)
11. Dierckens, K.E., Harrison, A.B., Leung, C.K., Pind, A.V.: A data science and engineering solution for fast k-means clustering of big data. In: IEEE BigDataSE 2017, pp. 925–932 (2017)
12. Han, J., Pei, J., Yin, Y.: Mining frequent patterns without candidate generation. In: ACM SIGMOD 2000, pp. 1–12 (2000)
13. Han, Z., Leung, C.K.: FIMaaS: scalable frequent itemset mining-as-a-service on cloud for non-expert miners. In: BigDAS 2015, pp. 84–91. ACM (2015) https://doi.org/10.1145/2837060.2837072
14. Jang, S., Chi, S., Yoo, K., Nasridinov, A.: Improving availability of manufacturing plant using association rule. In: BigDAS-L 2016 (2016)
15. Jiang, F., Leung, C.K., Sarumi, O.A., Zhang, C.Y.: Mining sequential patterns from uncertain big DNA data in the Spark framework. In: IEEE BIBM 2016, pp. 874–881 (2016)

16. Lakshmanan, L.V.S., Leung, C.K., Ng, R.T.: The segment support map: scalable mining of frequent itemsets. ACM SIGKDD Explorations 2(2), pp. 21–27 (2000)
17. Lee, R.C., Cuzzocrea, A., Lee, W., Leung, C.K.: Majority voting mechanism in interactive social network clustering. In: ACM WISM 2017, art. 14 (2017)
18. Leung, C.K.: Big data mining and computing in a smart world. In: IEEE UIC-ATC-ScalCom-CBDCom-IoP 2015, pp. xcviii (2015)
19. Leung, C.K.: Big data mining applications and services. In: BigDAS 2015, pp. 1–8. ACM (2015) https://doi.org/10.1145/2837060.2837076
20. Leung, C.K.: Data and visual analytics for emerging databases. In: EDB 2017. LNEE, vol. 461, pp. 203–213 (2017) https://doi.org/10.1007/978-981-10-6520-0_21
21. Leung, C.K.: Frequent itemset mining with constraints. Encyclopedia of Database Systems, 2e. Springer (2016) https://doi.org/10.1007/978-1-4899-7993-3_170-2
22. Leung, C.K.: Mathematical model for propagation of influence in a social network. Encyclopedia of Social Network Analysis and Mining, 2e. Springer (2017) https://doi.org/10.1007/978-1-4614-7163-9_110201-1
23. Leung, C.K.: Mining frequent itemsets from probabilistic datasets. In: EDB 2013, pp. 137–148 (2013)
24. Leung, C.K., Braun, P., Enkhee, M., Pazdor, A.G.M., Sarumi, O.A., Tran, K.: Knowledge discovery from big social key-value data. In: IEEE CIT 2016, pp. 484–491 (2016)
25. Leung, C.K., Jiang, F.: Efficient mining of 'following' patterns from very big but sparse social networks. In: IEEE/ACM ASONAM 2017, pp. 1025–1032. ACM (2017)
26. Leung, C.K., Jiang, F., Pazdor, A.G.M.: Bitwise parallel association rule mining for web page recommendation. In: IEEE/WIC/ACM WI 2017, pp. 662–669. ACM (2017)
27. Leung, C.K., Mateo, M.A.F., Brajczuk, D.A.: A tree-based approach for frequent pattern mining from uncertain data. In: PAKDD 2008. LNCS (LNAI), vol. 5012, pp. 653–661 (2008)
28. Leung, C.K., Tanbeer, S.K., Cameron, J.J.: Interactive discovery of influential friends from social networks. Social Network Analysis and Mining 4(1), art. 154. Springer (2014)
29. Renart, E.G., Diaz-Montes, J., Parashar, M.: Data-driven stream processing at the edge. In: IEEE ICFEC 2017, pp. 31–40 (2017)
30. Sajjad, H.P., Danniswara, K., Al-Shishtawy, A., Vlassov, V.: SpanEdge: towards unifying stream proccessing over central and near-the-edge data centers. In: IEEE/ACM SEC 2016, pp. 168–178. IEEE (2016)
31. Song, Y., Yau, S.S., Yu, R., Zhang, X., Xue, G.: An approach to QoS-based task distribution in edge computing networks for IoT applications. In: IEEE EDGE 2017, pp. 32–39 (2017)
32. Tanbeer, S.K., Ahmed, C.F., Jeong, B.-S.: Parallel and distributed frequent pattern mining in large databases. In: IEEE HPCC 2009, pp. 407–414 (2009)

Constrained Frequent Pattern Mining from Big Data Via Crowdsourcing

Calvin S. H. Hoi, Daniyal Khowaja and Carson K. Leung

Abstract High volumes of wide varieties of valuable data of different veracities can be easily generated or collected at a high velocity from various big data applications and services. Embedded in these big data are valuable knowledge and useful information, which can be discovered by data science solutions. As a popular data science task, frequent pattern mining aims to discover implicit, previously unknown and potentially useful information and valuable knowledge in terms of sets of frequently co-occurring items. Many of the existing frequent pattern mining algorithms return large numbers of frequent patterns, of which only a small portion may be of interest to users. In this paper, we present a constrained mining algorithm that allows crowds of users to collaboratively vote for their interesting patterns. Such an algorithm takes the benefits of crowdsourcing, crowdvoting and collaborative filtering for the data analytics and mining of popular constrained frequent patterns from big data applications and services.

Keywords Big data · Big data applications and services · Collaborative filtering
Crowdsourcing · Crowdvoting · Data analytics · Data mining · Data science
Frequent patterns · Knowledge discovery in databases

1 Introduction

As technology advances, high volumes of a wide variety of data of different veracities (e.g. uncertain data [31, 36]) can be easily generated or collected at a high velocity from rich sources of data in various real-life big data applications and services [27] such as bioinformatics [20], games [8, 9], smart worlds [26], social media [6, 11], social networks [25, 30, 32, 33, 37], Web [35] and wireless sensor systems [12]. Embedded in these big data are useful information and valuable knowledge. Hence,

C. S. H. Hoi · D. Khowaja · C. K. Leung (✉)
University of Manitoba, Winnipeg, MB, Canada
e-mail: kleung@cs.umanitoba.ca

W. Lee and C. K. Leung (eds.), *Big Data Applications and Services 2017*,
Advances in Intelligent Systems and Computing 770,
https://doi.org/10.1007/978-981-13-0695-2_9

data science [14]—which applies data mining, database, machine learning, visualization, mathematical and/or statistical techniques to big data for *data analytics* [28]—is in demand.

Among popular data science tasks, *frequent pattern mining* [19, 24] aims to discover sets of frequently co-occurring items. The mined patterns help data scientists understand consumer behaviour, know the web surfing patterns and get an insight about popular events. In other words, the mined patterns help the big data application and service providers to get better understanding of the behaviours of their consumers (i.e. big data application and service recipients or users).

Early frequent pattern mining algorithms (e.g. Apriori [1]) mostly execute the mining process in serial on a single local computer. To improve algorithmic efficiency, later algorithms perform the computation in distributed and parallel environments by using computer clusters or grids. As volumes of big data can go beyond the ability of commonly used software to capture, manage and process within a tolerable elapsed time, new forms of processing, mining and analysing data—such as MapReduce-based frequent pattern mining algorithms [7] that usually run in clouds—have been developed to enable enhanced decision-making, insight, knowledge discovery and process optimization. In general, cloud computing makes use of a network of remote servers hosted on the Internet to store, manage, process and mine the data.

Although many existing algorithms efficiently discover frequent patterns, most of them provide little or no support for user focus when mining data. In many real-life applications and services, the users may have some particular phenomena in mind on which to focus the mining (e.g. medical analysts may want to focus on only those sensor data belonging to patients suspected to suffer from asthma instead of all the patients, business analysts may want to focus on popular checkpoints for consumers of a certain age group instead of all their consumers, and application developers and service providers may want to focus on the applications or services that are interesting to certain consumers). Without user focus, the users often need to wait for a long period of time for numerous frequent patterns, out of which only a tiny fraction may be interesting to the users. This implies long and unnecessary computation. Hence, *constrained mining* [29]—which aims to discover those frequent patterns that satisfy the user-specified constraints—is needed. Several existing constrained frequent pattern mining algorithms [13, 23, 39] allow application developers and service providers (e.g. business analysts, business owners and store managers) to express their interests (e.g. for the purposes of promotional marketing or expanded services) by specifying constraints. However, there are situations in which the application users and service recipients (e.g. consumers) would also like to express their interests. Inspired by the idea of *crowdsourcing* [16, 41] and *collaborative filtering* [34], we conduct—in this paper—a feasibility study on constrained frequent pattern mining from big data via crowdsourcing. Here, we also design a constrained frequent pattern mining algorithm that allows crowds of users to express their interests by specifying their constraints and crowdvoting their choices.

Our *key contribution* of this paper includes our algorithm for mining frequent patterns that satisfy user-specified constraints for big data applications and services via crowdsourcing. Evaluation results show that our algorithm can be considered as a practical alternative for efficient constrained mining of patterns.

The remainder of this paper is organized as follows. Next section reviews some related works. Section 3 presents our algorithm for constrained pattern mining from big data with crowdsourcing. Section 4 explains our evaluation. Finally, conclusions are drawn in Sect. 5.

2 Related Works

Frequent pattern mining aims to discover sets of frequently co-occurring items. Many frequent pattern mining algorithms were Apriori-based [1], which rely on a generate-and-test paradigm to discover frequent patterns from transaction datasets by first generating candidates and then checking their actual frequency (i.e. occurrences) against the datasets. To improve algorithmic efficiency, other serial frequent pattern mining algorithms (e.g. tree-based algorithms [17]) have been developed. In addition, there are also distributed and parallel algorithms [35, 40]. They find frequent patterns by using multiple processors that have either (a) access to a shared memory with which information is exchanged among processors or (b) their own private memory (i.e. distributed memory) with which for information is exchanged via message passing among the processors. These algorithms are mainly implemented using Open Multi-Processing (OpenMP) or Message Passing Interface (MPI) and run in computer clusters or grids. Cluster computing usually involves a group of distributed or parallel computers that are interconnected through high-speed networks such as local area networks; grid computing coordinates heterogeneous networked loosely coupled computers.

To handle big data, MapReduce-based frequent pattern mining algorithms [7] read input data, divide them into several partitions (sub-problems) and assign them to different processors. Each processor executes the 'map' function on each partition by taking a pair of $\langle key_1, value_1\rangle$ and returning a list of $\langle key_2, value_2\rangle$-pairs as an intermediate result, which is then shuffled and sorted. Each processor then executes the 'reduce' function on a single key_2 from this intermediate result by combining, aggregating, summarizing, filtering and/or transforming the list of values associated with a given key_2 (for all keys). As a result, a single (aggregated or summarized) $value_3$ is returned. Many of these algorithms are run in (public, private or hybrid) clouds. Cloud computing [4] involves a group of interconnected and virtualized computers to provide on-demand services such as function-as-a-service (FaaS), infrastructure-as-a-service (IaaS), network-as-a-service (NaaS), platform-as-a-service (PaaS), software-as-a-service (SaaS) and mining-as-a-service (MaaS) [18].

To focus the mining, ***constrained frequent pattern mining*** algorithms [13, 23, 29] allow application developers and service providers, as well as data scientists, to use a rich set of SQL-style constraints to specify their interest in finding only those frequently occurring sets of market basket items that satisfy the specified constraints. This avoids unnecessary computation for mining those uninteresting frequent itemsets. Besides market basket items, the set of constraints can also be imposed on items, events or objects in other domains. However, many existing constrained frequent pattern mining algorithms were designed for handling constraints specified by *service providers*. In contrast, our algorithm is designed to handle constraints specified by *consumers* and *service recipients* in addition to the application developers and service providers.

Partially due to its diversity, availability and efficiency, ***crowdsourcing***—which 'outsources' some specific tasks to crowds of public users who collaboratively contribute to the tasks—has become an emerging area. Sample categories of crowdsourcing include the following:

- Crowdvoting, in which the crowds express their opinions or judgements on certain topics via online votes.
- Crowdsolving, in which the crowds collectively help solve some specific problems, macro-tasks or micro-tasks via their collective intelligence or collective problem-solving skills. For example, Wikipedia[1] is a product of crowdsolving, in which the crowds collectively contribute to different encyclopedia entries.
- Crowdsearch, in which the crowds collaboratively help search and find some lost items or missing persons.
- Crowdfunding, in which the crowds cumulatively contribute a small amount of fund towards some monetary goals.

Through these categories of social collaboration-based activities, crowdsourcing can be applicable in various big data applications and services by performing data science tasks like unsupervised learning (clustering), semi-supervised learning and supervised learning (classification) [16], as well as frequent pattern and association rule mining [3].

In general, frequent patterns capture the sets of frequently co-occurring merchandise items or co-located events. *Association rules*—each usually is of the form $A \Rightarrow C$ where the antecedent A and the consequence C of a rule are frequent patterns—reveal interesting relationships among these frequent patterns. ***Interestingness measures*** [15] capture the following aspects of the association rules and their related frequent patterns:

- objective aspects by measuring conciseness of summaries, coverage (or generality), diversity of summaries (via the use of measures like variance), peculiarity of cells in summaries or reliability;
- semantics aspects by measuring actionability (or applicability) or utility (via the use of measures like weighted support and share) of the rules or patterns;
- subjective aspects by measuring novelty or surprisingness (or unexpectedness).

The following are some commonly used probability-based objective interestingness measures [15] for association rules of the form $A \Rightarrow C$:

1. support($A \Rightarrow C$) = $P(A \cup C)$,
2. confidence($A \Rightarrow C$) = $P(C|A) = \frac{P(A \cup C)}{P(A)}$,

[1] https://en.wikipedia.org/wiki/Main_Page.

3. coverage($A \Rightarrow C$) = $P(A)$,
4. lift($A \Rightarrow C$) = $\frac{P(C|A)}{P(C)} = \frac{P(A \cup C)}{P(A)P(C)}$,
5. linear correlation coefficient($A \Rightarrow C$) = $\frac{P(A \cup C) - P(A)P(C)}{\sqrt{P(A)P(C)P(\neg A)P(\neg C)}}$ and
6. Piatetsky-Shapiro($A \Rightarrow C$) = $P(A \cup C) - P(A)P(C)$.

To mine (unconstrained) association rules from crowds, CrowdMiner [2, 3] uses active learning techniques for producing closed questions (e.g. whether '$A \Rightarrow C$' is interesting?) or open questions (e.g. '? ⇒ ?') to interactively ask the crowds. It uses support and confidence as interestingness measures. In contrast, our algorithm mines *constrained* frequent patterns and association rules. Moreover, our algorithm allows the crowds to select their interesting items. Hence, it uses the crowds' preference in these selections as the interestingness measure.

3 Crowdsourcing-Based Constrained Mining

In this section, we describe our crowdsourcing-based constrained frequent pattern mining algorithm called *Crowd-CFP-Miner*. Our algorithm allows both the service providers and the consumers to express their interests by specifying constraints. More specifically, similar to some existing constrained frequent pattern mining algorithms, our algorithm allows the application developers, service providers and/or data scientists to express their interests on domain items. For example, the algorithm allows the merchants to express their interests (say, for product promotions) in focusing the mining on only some selected merchandise items (say, laptops). As another example, the algorithm allows the merchants to express their interests (say, for bundling) in focusing the mining on some targeted merchandise items and their related items. The algorithm handles these types of service providers' constraints by (i) focusing exclusively on only the targeted items (and excluding other items) or (ii) focusing on the mandatory items (i.e. targeted items) together with other optional items.

Unlike the existing constrained frequent pattern mining algorithms, our algorithm allows the service providers and/or data scientists to express their interests on targeted crowds. For example, the algorithm allows medical analysts to express their interests (say, for analysis on medical treatment or drug efficacy) in focusing the mining on only crowds of patients who suffer from asthma. As an example, the algorithm also allows business analysts to express their interests (say, for promotion of youth tours or student travel package) in focusing the mining on only crowds of travellers of a certain age group (say, youth) or occupation (say, students). The algorithm handles these types of service providers' constraints by 'outsourcing' the frequent pattern mining task to only *targeted crowds* of public users—i.e. crowds of users who satisfy the constraints (e.g. youth, students and asthma patients)—based on the users' profile (say, on age and occupation) or users' voluntary information (say, asking whether or not the user is suffered from asthma).

In addition to incorporating the service providers' constraints, our algorithm also incorporates constraints specified by the crowds (i.e. consumers, service recipients).

More specifically, the algorithm does not require or expect crowds of public users to be able to express their interests by technically specifying SQL-style constraints (cf. data scientists are expected to technically specify SQL-style constraints when using many of the existing constrained frequent pattern mining algorithms). Instead, our algorithm displays some (relevant) or all of the domain items for the crowds to choose from. For example, the algorithm allows the crowds to select some of the targeted items specified by the service providers. By doing so, the crowds can choose a subset of interested items (which are interesting to the crowds) from the targeted items (which are interesting to the service providers) so as to enhance the mining process by focusing on items that are mutually interested by both parties.

After incorporating the constraints specified by both the service providers and crowds, our algorithm starts the crowdsourcing-based mining process. More specifically, the algorithm lists the items (that are mutually interested by both parties), from which the crowds can choose popular or frequent items. Here, the algorithm provides the following two options or two levels of granularity:

1. Boolean choice: The crowds are asked to choose the items that they think are popular or frequent.
2. Categorical choice: The crowds are asked to categorize the items into several categories (e.g. never purchased, rarely purchased, occasionally purchased, often purchased, regularly purchased, etc.) based on the crowds' expected or perceived popularity or frequency of the displayed items.

As a variant to listing the items (that are mutually interested by both the service providers and crowds), the algorithm first applies the traditional frequent pattern mining algorithm to find the frequent singletons (i.e. frequent 1-itemsets) by using those items that are mutually interested by both parties. The algorithm then asks the crowds to choose—from this collection of frequent 1-itemsets—the items that are perceived to be popular or frequent.

Next, our algorithm lists the items (that are mutually interested by both parties), from which the crowds can choose sets of popular or frequent items that are purchased together. Again, the algorithm provides the following two options or two levels of granularity:

1. Boolean choice: The crowds are asked to choose collections of items that they think are popular or frequent.
2. Categorical choice: The crowds are asked to categorize collections of items into several categories (e.g. never purchased together, rarely purchased together, occasionally purchased together, often purchased together, regularly purchased together, etc.) based on the crowds' expected or perceived popularity or frequency of the collections of displayed items.

Again, as a variant to listing the collections of items (that are mutually interested by both the service providers and crowds), the algorithm first applies the traditional frequent pattern mining algorithm to find the frequent non-singletons (i.e. frequent k-itemsets where $k \geq 2$) by using (i) those items that are mutually interested by both parties or (ii) those chosen 1-itemsets. The algorithm then asks the crowds to choose—from these collections of frequent k-itemsets—the k-itemsets that are perceived to be popular or frequent.

4 Evaluation

To evaluate our algorithm, we conducted two case studies using the following transactional datasets:

1. An online retail dataset[2] [10] from the UCI Machine Learning Repository: This dataset captures 541,909 transactions occurring between 1 December 2010 and 9 December 2011 inclusive for a UK-based and registered non-store online retail.
2. A store sales dataset: It captures 9994 transactions occurring during 2014 to 2017 inclusive from different branches of a multi-city store in the USA.

In the first case study, we—as data scientists—specified a domain-item constraint 'item category = gift' to express our interests in focusing on only gift items. By incorporating this constraint, our algorithm focuses the mining on about 4000 unique gift items occurring in 25,900 transactions (cf. all online retail items in 541,909 transactions) in the online retail dataset. In addition, we also specified a targeted-crowd constraint 'age $\geq$ 18' to express our interests in focusing on the targeted crowds of public users whose ages are 18 or above.

After incorporating the constraints from service providers or data scientists, our algorithm then 'outsourced' the mining task to the targeted crowds—whose demographics information can be found in Fig. 1—by asking the crowds to participate in an online survey posted on SurveyMonkey.[3] Here, *Amazon Mechanical Turk* (*MTurk*)[4] was used as a crowdsourcing platform to enable the crowds and service providers (or data scientists) in coordinating the use of human collective intelligence to perform a crowdsourcing task (e.g. constrained mining task for our case study). MTurk runs on a requester–worker model, in which requesters post a task (e.g. constrained mining task) and the workers complete the task by answering some visual questions [5], article feedback [21], performance evaluation [22] and/or design critiques [38]. In this case study, we requested the workers (i.e. crowds) to express their preference or interested items among the service providers' targeted items (i.e. gift items) by answering some visual questions like 'Which of the following items you would buy?'. Examples of interested items include teacups, saucers, and 3-tier cake stands.

Afterwards, our algorithm showed the crowds with a list of gifts items, from which the crowds were asked to select some individual items that are perceived to be frequently purchased by consumers. Recall that, as our algorithm provides two variants, such a list can be drawn from either one of the following:

[2]http://archive.ics.uci.edu/ml/datasets/online+retail.

[3]https://www.surveymonkey.com/.

[4]https://www.mturk.com/.

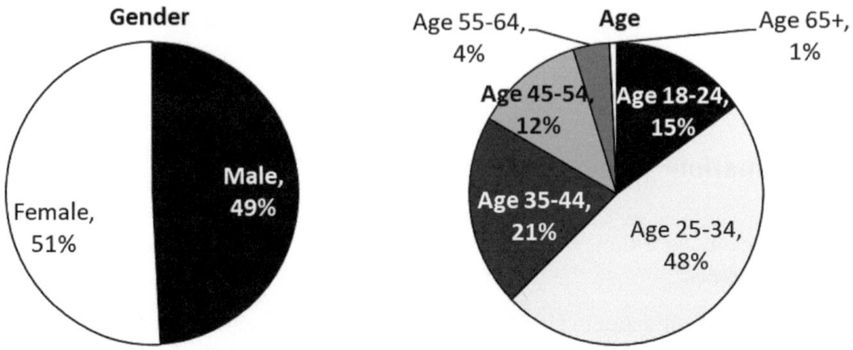

Fig. 1 Demographics information about the targeted crowds

- a collection of items that are mutually interested by both service providers and the crowds (i.e. crowd-preferred, targeted domain items) or
- a collection of 1-itemsets (i.e. frequent singletons) that are returned by traditional unconstrained frequent pattern mining algorithm.

When asked to select some individual items, the crowds were given one or both of the following types of choices:

- Boolean choices with questions like 'In your opinion, which of the following items are frequently purchased by consumers?', and/or
- Categorical choices with questions like 'In your opinion, how frequently are the following items purchased by consumers: (a) never, (b) rarely, (c) occasionally, (d) often or (e) always?'.

Along this direction, our algorithm then showed the crowds with a list of gifts items, from which the crowds were asked to select some sets of items that are expected to be frequently purchased together by consumers in the form of 'Given a set A of items, select items (to be put in a set C) that are expected to be frequently purchased together with A for an association rule $A \Rightarrow C$'. Again, due to the two variants supported by our algorithm, such a list can be drawn from either one of the following:

- a collection of items that are mutually interested by both service providers and the crowds (i.e. crowd-preferred, targeted domain items) or
- a collection of k-itemsets (i.e. frequent non-singletons) that are returned by traditional frequent pattern mining algorithm (where $k \geq 2$) using interestingness measures like support, confidence and lift described in Sect. 2.

When asked to select some items, the crowds were given one or both of the following types of choices:

- Boolean choices with questions like 'In your opinion, given a set A of items, which of the following items are frequently purchased together with A by consumers?', and/or

– Categorical choices with questions like 'In your opinion, given a set A of items, how frequently are the following set C of items purchased together by consumers: (a) never, (b) rarely, (c) occasionally, (d) often or (e) always?'.

Similarly, our second case study was conducted in a similar fashion, in which we—as data scientists—specified a domain-item constraint 'item category = office supplies' to express our interests in focusing on only office supplies (e.g. pencil sharpener, calculators, and inkjet printers) among all the store items in the dataset. By incorporating this constraint, our algorithm focuses the mining on about 1263 unique office supplies occurring in 5496 transactions (cf. all store items in 9994 transactions) in the multi-city store sales dataset. Again, we also specified a targeted-crowd constraint 'age $\geq$ 18' to express our interests in focusing on the targeted crowds of public users whose ages are 18 or above, to which our algorithm 'outsourced' the mining task. The crowds were asked to specify their preference and select k-itemsets (where $k \geq 1$) that were expected to be purchased together.

From these two case studies, we made many observations. The following are some interesting ones:

1. The crowds tended to choose more itemsets from categorical choices than Boolean choices. This is because the crowds interpreted 'frequent' itemsets in the Boolean choices as those items that were 'often' or 'always purchased together' in the categorical choices (and thus ignored those items that were 'occasionally' or 'rarely purchased together' in the categorical choices).
2. When our algorithm chose itemsets from a collection of crowd-preferred, targeted domain items, it enabled crowdvoting. As the chosen itemsets truly reflected the crowds' perceived frequency and popularity on these itemsets, our algorithm serves as a good alternative to existing constrained frequent pattern mining algorithms.
3. When our algorithm chose itemsets from a collection of frequent itemsets returned by traditional frequent pattern mining algorithms, it enabled the users to further focus the mining on some manageable subsets of frequent patterns.

All these observations demonstrated the practicality of our algorithm.

5 Conclusions

In this paper, we presented an algorithm for constrained mining of frequent patterns from big data via crowdsourcing. The algorithm allows both (i) the application and service providers and (ii) crowds of consumers to express their interests and preference on items that form frequent patterns and their related association rules. The mined results help reveal useful information and valuable knowledge about the behaviours of crowds of consumers. Our two feasibility case studies demonstrated the practicality of our algorithm in crowdsourcing-based constrained frequent pattern mining from big data. As ongoing work, we continue to conduct more extensive evaluation and explore more functional features to be provided to the users.

Acknowledgements This project is partially supported by (i) Natural Sciences and Engineering Research Council of Canada (NSERC) and (ii) University of Manitoba.

References

1. Agrawal, R., Srikant, R.: Fast algorithms for mining association rules in large databases. In: VLDB 1994, pp. 487–499. Morgan Kaufmann (1994)
2. Amsterdamer, Y., Grossman, Y., Milo, T., Senellart, P.: Crowd mining. In: ACM SIGMOD 2013, pp. 241–252 (2013)
3. Amsterdamer, Y., Grossman, Y., Milo, T., Senellart, P.: CrowdMiner: mining association rules from the crowd. PVLDB 6(12), pp. 1250–1253 (2013)
4. Bierzynski, K., Escobar, A., Eberl, M.: Cloud, fog and edge: cooperation for the future? In: FMEC 2017, pp. 62–67. IEEE (2017)
5. Bigham, J.P., Jayant, C., Ji, H., Little, G., Miller, A., Miller, R.C., Miller, R., Tatarowicz, A., White, B., White, S., Yeh, T.: VizWiz: nearly real-time answers to visual questions. In: ACM UIST 2010, pp. 333–342 (2010)
6. Braun, P., Cuzzocrea, A., Doan, L.M.V., Kim, S., Leung, C.K., Matundan, J.F.A., Singh, R.R.: Enhanced prediction of user-preferred YouTube videos based on cleaned viewing pattern history. Procedia Computer Science 112, pp. 2230–2239. Elsevier (2017)
7. Braun, P., Cuzzocrea, A., Jiang, F., Leung, C.K., Pazdor, A.G.M.: MapReduce-based complex big data analytics over uncertain and imprecise social networks. In: DaWaK 2017. LNCS, vol. 10440, pp. 130–145 (2017)
8. Braun, P., Cuzzocrea, A., Keding, T.D., Leung, C.K., Pazdor, A.G.M., Sayson, D.: Game data mining: clustering and visualization of online game data in cyber-physical worlds. Procedia Computer Science 112, pp. 2259–2268. Elsevier (2017)
9. Brown, J.A., Cuzzocrea, A., Kresta, M., Kristjanson, K.D.L., Leung, C.K., Tebinka, T.W.: A machine learning system for supporting advanced knowledge discovery from chess game data. In: IEEE ICMLA 2017, pp. 649–654 (2017)
10. Chen, D., Sain, S.L., Guo, K.: Data mining for the online retail industry: a case study of RFM model-based customer segmentation using data mining. Journal of Database Marketing and Customer Strategy Management 19(3), pp. 197–208 (2012)
11. Choudhery, D., Leung, C.K.: Social media mining: prediction of box office revenue. In: IDEAS 2017, pp. 20–29. ACM (2017)
12. Cuzzocrea, A., Grasso, G.M., Jiang, F., Leung, C.K.: Mining uplink-downlink user association in wireless heterogeneous networks. In: IDEAL 2016. LNCS, vol. 9937, pp. 533–541 (2016)
13. Cuzzocrea, A., Leung, C.K., MacKinnon, R.K.: Mining constrained frequent itemsets from distributed uncertain data. Future Generation Computer Systems 37, pp. 117–126. Elsevier (2014)
14. Dierckens, K.E., Harrison, A.B., Leung, C.K., Pind, A.V.: A data science and engineering solution for fast k-means clustering of big data. In: IEEE BigDataSE 2017, pp. 925–932 (2017)
15. Geng, L., Hamilton, H.J.: Interestingness measures for data mining: a survey. ACM CSUR 38(3), art. 9 (2006)
16. Guo, X., Wang, H., Song, Y., Hong, G.: Brief survey of crowdsourcing for data mining. ESWA 41(17), pp. 7987–7994. Elsevier (2014)
17. Han, J., Pei, J., Yin, Y.: Mining frequent patterns without candidate generation. In: ACM SIGMOD 2000, pp. 1–12 (2000)
18. Han, Z., Leung, C.K.: FIMaaS: scalable frequent itemset mining-as-a-service on cloud for non-expert miners. In: BigDAS 2015, pp. 84–91. ACM (2015) https://doi.org/10.1145/2837060.2837072
19. Jang, S., Chi, S., Yoo, K., Nasridinov, A.: Improving availability of manufacturing plant using association rule. In: BigDAS-L 2016 (2016)

20. Jiang, F., Leung, C.K., Sarumi, O.A., Zhang, C.Y.: Mining sequential patterns from uncertain big DNA data in the Spark framework. In: IEEE BIBM 2016, pp. 874–881 (2016)
21. Kittur, A., Chi, E.H., Suh, B.: Crowdsourcing user studies with Mechanical Turk. In: ACM CHI 2008, pp. 453–456 (2008)
22. Komarov, S., Reinecke, K., Gajos, K.Z.: Crowdsourcing performance evaluations of user interfaces. In: ACM CHI 2013, pp. 207–216 (2013)
23. Lakshmanan, L.V.S., Leung, C.K., Ng, R.T.: Efficient dynamic mining of constrained frequent sets. ACM TODS 28(4), pp. 337–389 (2003)
24. Lakshmanan, L.V.S., Leung, C.K., Ng, R.T.: The segment support map: scalable mining of frequent itemsets. ACM SIGKDD Explorations 2(2), pp. 21–27 (2000)
25. Lee, R.C., Cuzzocrea, A., Lee, W., Leung, C.K.: Majority voting mechanism in interactive social network clustering. In: ACM WISM 2017, art. 14 (2017)
26. Leung, C.K.: Big data mining and computing in a smart world. In: IEEE UIC-ATC-ScalCom-CBDCom-IoP 2015, pp. xcviii (2015)
27. Leung, C.K.: Big data mining applications and services. In: BigDAS 2015, pp. 1–8. ACM (2015) https://doi.org/10.1145/2837060.2837076
28. Leung, C.K.: Data and visual analytics for emerging databases. In: EDB 2017. LNEE, vol. 461, pp. 203–213 (2017) https://doi.org/10.1007/978-981-10-6520-0_21
29. Leung, C.K.: Frequent itemset mining with constraints. Encyclopedia of Database Systems, 2e. Springer (2016) https://doi.org/10.1007/978-1-4899-7993-3_170-2
30. Leung, C.K.: Mathematical model for propagation of influence in a social network. Encyclopedia of Social Network Analysis and Mining, 2e. Springer (2017) https://doi.org/10.1007/978-1-4614-7163-9_110201-1
31. Leung, C.K.: Mining frequent itemsets from probabilistic datasets. In: EDB 2013, pp. 137–148 (2013)
32. Leung, C.K., Braun, P., Enkhee, M., Pazdor, A.G.M., Sarumi, O.A., Tran, K.: Knowledge discovery from big social key-value data. In: IEEE CIT 2016, pp. 484–491 (2016)
33. Leung, C.K., Jiang, F.: Efficient mining of 'following' patterns from very big but sparse social networks. In: IEEE/ACM ASONAM 2017, pp. 1025–1032. ACM (2017)
34. Leung, C.K., Jiang, F., Dela Cruz, E.M., Elango, V.S.: Association rule mining in collaborative filtering. In: Collaborative Filtering Using Data Mining and Analysis, pp. 159–179 (2017)
35. Leung, C.K., Jiang, F., Pazdor, A.G.M.: Bitwise parallel association rule mining for web page recommendation. In: IEEE/WIC/ACM WI 2017, pp. 662–669. ACM (2017)
36. Leung, C.K., Mateo, M.A.F., Brajczuk, D.A.: A tree-based approach for frequent pattern mining from uncertain data. In: PAKDD 2008. LNCS (LNAI), vol. 5012, pp. 653–661 (2008)
37. Leung, C.K., Tanbeer, S.K., Cameron, J.J.: Interactive discovery of influential friends from social networks. Social Network Analysis and Mining 4(1), art. 154. Springer (2014)
38. Luther, K., Tolentino, J., Wu, W., Pavel, A., Bailey, B.P., Agrawala, M., Hartmann, B., Dow, S.P.: Structuring, aggregating, and evaluating crowdsourced design critique. In: ACM CSCW 2015, pp. 473–485 (2015)
39. Rahman, M.M., Ahmed, C.F., Leung, C.K., Pazdor, A.G.M.: Frequent sequence mining with weight constraints in uncertain databases. In: ACM IMCOM 2018, art. 48 (2018) https://doi.org/10.1145/3164541.3164627
40. Tanbeer, S.K., Ahmed, C.F., Jeong, B.-S.: Parallel and distributed frequent pattern mining in large databases. In: IEEE HPCC 2009, pp. 407–414 (2009)
41. Yang, H.C., Lee, C.H.: Toward crowdsourcing data mining. In: IDAM 2013, pp. 107–110. Springer (2013)

Data-Driven Prediction of Ship Destinations in the Harbor Area Using Deep Learning

Kwang Il Kim and Keon Myung Lee

Abstract The prediction of ship destinations in the harbor can be utilized to identify future routes for navigating ships. The maritime traffic data are broadly classified into the ship trajectory data and the port information management data. These data have been accumulated for many years on the shore base station of different agencies, and are being utilized for evaluation of collision risk, prediction of vessel traffic, and other maritime statistical analysis. This paper presents a new destination prediction model of navigating ships in the harbor which consists of the candidate harbor proposal module and the position–direction filter module. The candidate harbor proposal module is trained by a deep neural network which makes use of the characteristics of ships and the occupancy distributions of piers. The position–direction filter module leaves out non-promising ones from the harbor list provided by the candidate proposal module, with respect to the current position and direction of navigating ship. In the experiments on real vessel traffic data, the proposed method has shown that its accuracy is higher than the frequency-based baseline method by about 10–15%.

Keywords Maritime data analysis · Deep neural network · Prediction model
Vessel traffic · Harbor proposal module · Position–direction filter

1 Introduction

The vessel traffic in most harbor areas keeps increasing and the concerns of traffic accidents and effective traffic management also keep increasing. The vessel traffic service (VTS) officers and ship navigators have to carry out stressful missions especially in harbor areas for the safety and efficiency of vessel and harbor traffic. The

K. I. Kim · K. M. Lee (✉)
Department of Computer Science, Chungbuk National University,
Cheongju, South Korea
e-mail: kmlee@cbnu.ac.kr

K. I. Kim
e-mail: kikim82@cbnu.ac.kr

W. Lee and C. K. Leung (eds.), *Big Data Applications and Services 2017*,
Advances in Intelligent Systems and Computing 770,
https://doi.org/10.1007/978-981-13-0695-2_10

advances in sensor technology and communication infrastructure have made various maritime traffic data for vessels and harbors collected and analyzed. Historical data of vessel traffic are valuable assets from which important clues can be extracted for the enhancement of security and efficiency in harbor traffic [1].

Large-scale vessels and high-end vessels are well equipped with multiple communication channels including digital and analog devices, but some small-scale vessels like fish boats have limited communication channels. Although the VTS station in a harbor monitors and communicates with vessels in its duty area, it is hardly possible for the VTS to collect all traffic information in real time. One of the important tasks in VTS and navigating vessels is to predict the future routes of vessels in the harbor area [2]. Once the destination of a vessel is identified, it is possible to predict its route.

The paper is concerned with a destination prediction model which is built from the maritime traffic data such as ship trajectory data and port information management system data. The existing methods use some simple statistical analysis like frequency distribution of destinations and also rely on the expertise of VTS officers and ship navigators. The proposed method uses a deep neural network (DNN) which is a machine learning model that builds a model from a collection of data. The proposed destination prediction model consists of the candidate harbor proposal module and the position–direction filter module. The candidate harbor proposal module is trained by a deep neural network which makes use of the characteristics of ships and the occupancy distributions of piers. The position–direction filter module leaves out non-promising ones from the harbor list provided by the candidate proposal module, with respect to the current position and direction of navigating ship.

The remainder of this paper is organized as follows: Sect. 2 reviews some related work. Section 3 describes the maritime traffic data used for destination prediction in the harbor area and presents the proposed prediction model for ship destinations. Section 4 shows some experiment results, and Sect. 5 draws the conclusions.

2 Related Work

There have been several works to predict ship routes in the coastal water area. They use the automatic identification system (AIS) data which contain the information about the vessels and their movements. Rhodes et al. [3] proposed a neural network-based model to predict the short-term motions of ships by training the normal behavior of vessels with AIS data. Pallotta et al. [4] developed a method to search the waypoints of vessels using a point-based incremental algorithm with the sliding time window, and to predict historical waypoints along with the posterior probabilities. Ristic et al. [5] proposed a motion prediction model which extracts the normal behavior patterns from sequential AIS data and compares the current trajectory with the extracted patterns. Those methods paid attention to the vessel motion prediction not in the harbor area but the coastal water area.

The traffic conditions in the harbor area are much more complicated than in the coastal water area, in terms of crowdedness of traffic, openness of area, geographical factors, pier facilities, and occupancies. Hence, AIS data are alone not sufficient to predict the vessel movements in the harbor area. Once the destination of a ship is known, it is relatively easy to predict its future route due to the regulations enforced by the harbor authorities. Kim et al. [6] proposed a statistical method to predict the destination of ships. The method uses the PORT-MIS (port management information system) data which contains the information for handling port operations. It counts the frequencies of destinations for each vessel from the PORT-MIS data and recommends the top 3 frequent destinations. In their experiments, the accuracy of the top frequent destination was less than 50%.

In the harbor area, it is important to predict the future movements of vessels for safety and efficiency. Either only AIS data or PORT-MIS data are not sufficient to predict the future movements as shown in the previous works. This paper is concerned with a destination prediction model that use both AIS data and PORT-MIS data together. We propose a deep learning based method to generate candidate destinations from characteristics of a vessel and its current movement. Deep learning is a machine learning technique that uses a deep neural network and has shown excellent performances in various difficult tasks. The multiple layers of a deep neural network allow the model to find useful and latent features for identifying complicated relationships between input and output spaces.

3 The Proposed Destination Prediction Method

3.1 The Maritime Traffic Data for the Prediction Model

The shore base stations like VTS stations and PORT-MIS system have collected a large amount of data about ship trajectories and port operations for a long time [7]. The ship trajectory data are collected from AIS data of which items can be categorized into dynamic, static, and voyage fields. The dynamic fields contain items which may change over time, such as ship position, speed, course, heading direction, and others. The static fields describe the static characteristics of ships, such as ship name, type, length, width, and draft information. The voyage fields contain data about a specific voyage of a ship, such as destination port and estimated time of arrival. The harbor operations are usually managed by an information system such as PORT-MIS which is a port management information system managed by a governmental body in South Korea. Table 1 shows data about pier/cargo operations and their related operations collected in PORT-MIS [8].

The proposed model uses both AIS data and pier operations data for a deep neural network training. Those data are managed in different databases in separated systems. The training data are prepared by integrating those related different databases into a single database as follows: The table of dynamic fields from the AIS system is

Table 1 Data available from PORT-MIS

Information type	List of information
Pier operation	Ship arrival/departure and berth operation
Cargo control	Cargo loading/unloading operation
Billing	Levy data generation, levy calculation, bill publication, billing error/arrear management
Statistics	Ship information, pier information, cargo, and pier operation statistics

regarded as the reference table and is joined with the PORT-MIS tables with respect to MMSI ID and time attributes. For the resulting table, the fields of ship entry and departure to and from piers in the port operation table are joined with AIS tables with respect to CallSign attribute. The integrated table is used as the training data from which input and output attributes are selected as follows:

Input attributes

- Number of ships berthing all port piers when the target data ship arrives;
- Ship information: ship tonnage, depth, length, type, and sailing area; and
- Ship trajectory data: ship positions, course, and speed.

Output attribute

- The target ship destination of port pier.

3.2 The Vessel Destination Prediction Model

The proposed vessel destination prediction model uses both AIS data and PORT-MIS data together to improve the accuracy of the model. The model consists of the candidate harbor proposal module and the position–direction filter module as shown in Fig. 1. The candidate harbor proposal module produces some candidate destinations from the abovementioned input attributes. Then, the position–direction filter module receives the candidate destinations and evaluates them with respect to their compatibility with the current position and navigation direction of a vessel. The module makes use of real-time data of the vessel. The candidate destinations which pass the examination of the position–direction filter module are given as the output of the destination prediction model. The maximum number of final destinations generated by the model is limited to the specified number, e.g., 5.

The candidate harbor proposal module is implemented with a deep neural network which is a multilayer Perceptron with six hidden layers, as shown in Fig. 2. In the network, the nodes from adjacent layers are fully connected and the ReLU (rectified linear unit) is used as the activation function. The final layer is the softmax layer that generates the probability-like outputs in which each node produces nonnegative

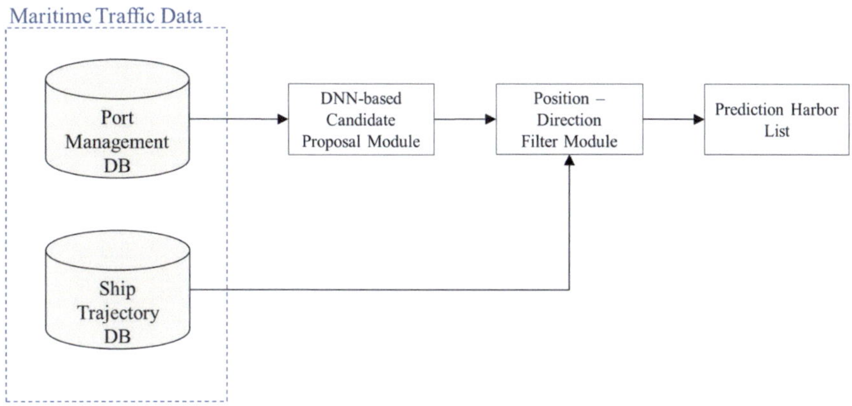

Fig. 1 Components of the proposed destination prediction model

values summing to 1. In the input layer, the following items are received as the input: sailing area, ship tonnage, ship depth, ship length, ship type, and pier occupation rate. Except pier occupation rate, all other input items are preprocessed to have categorical values. For example, the sailing area is expressed as having one of *inland*, *coastal*, and *ocean*. Those categorical attributes are encoded into one-hot vector representation. If an attribute can have three different values, the attribute is expressed in a one-hot vector with three elements like (1, 0, 0). Sailing area, ship tonnage, ship depth, ship length, and ship types have 3, 10, 10, 10, and 55 different values, respectively. Pier occupation rate is a numeric attribute with continuous domain [0, 1], in which the rate is computed by dividing the number of vessels at the moment with the maximum number of vessels berthed in the port. In the output layer, that is, the softmax layer, each node corresponds to a port in the harbor. Figure 2 shows the model for a harbor with 353 piers. In the model, the number of output node is 353 and the output of each output node is the prediction score. The higher the output value, the higher the probability with which the target vessel travels to the corresponding pier.

The loss function in the model training is the cross-entropy between the target output in one-hot representation and the network output. The model is trained with the error backpropagation algorithm which is the widely used training algorithm for neural network. Dropout is also used for avoiding the overfitting. The prepared data according to the procedure described in Sect. 3.1 are used as the training data for the model.

The position–direction filter module leaves out non-promising ones from the harbor list calculated by the candidate proposal module. The filter applied this module is generated using the vessel position, course, and the route gateline information. As shown in Fig. 3, the list of piers in the direction of the inbound vessel is selected as candidate piers.

The ship traffic routes in the harbor area are divided into several route gate lines. Each gate line stores a range of gate line position, input and output direction courses, and adjacent piers. Using these route gate line data, the current gate line and the sailing

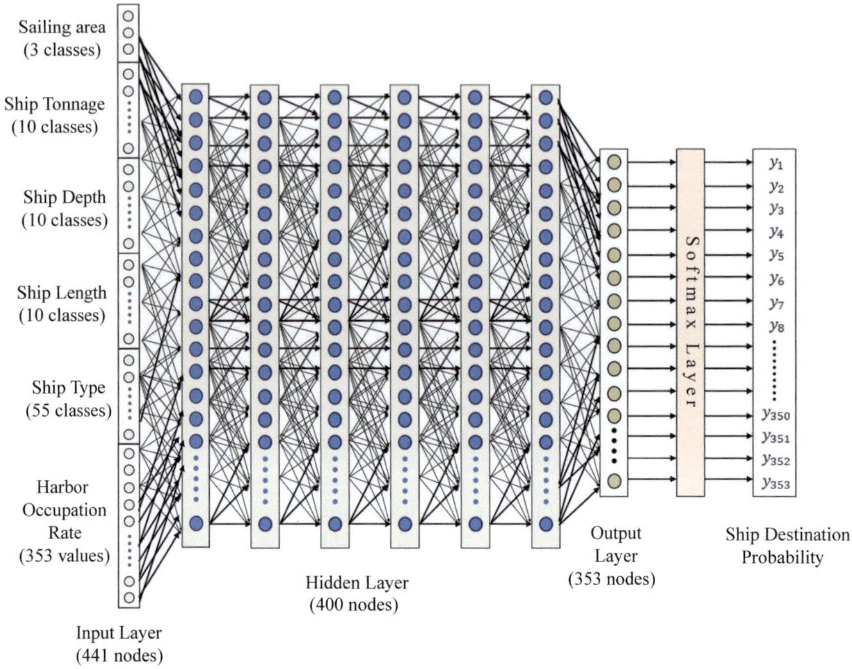

Fig. 2 The deep neural network for the candidate harbor proposal module

direction of the target vessel are selected by comparing the position and course of the target ship to the gate line information. The gate line candidates are extracted using the current gate line and the sailing direction, and these produce the list of piers in the direction of the vessel. This module generates the position–direction filter vector which consists of elements 0 and 1 by determining whether it is included in the list of piers. By multiplying the position–direction filter and the probability of the harbor candidates, non-promising piers are excluded. In Fig. 4, we present the pseudocode for the position–direction filter module.

4 Experiments

In order to evaluate the proposed method, we have implemented the method and applied it the real dataset. The deep neural network model of Fig. 2 is implemented with TensorFlow. As the training data, we collected the dataset for ship trajectories, ship entry and departure logs, and ship registration data of Yeosu Harbor, South Korea for 2 years (2014–2015). In the dataset, there are 254,700 ship entry and departure transactions for 353 pier and anchorages in the harbor. Among them, 229,230 records were used for training of the deep neural network model, and the remaining were used for testing.

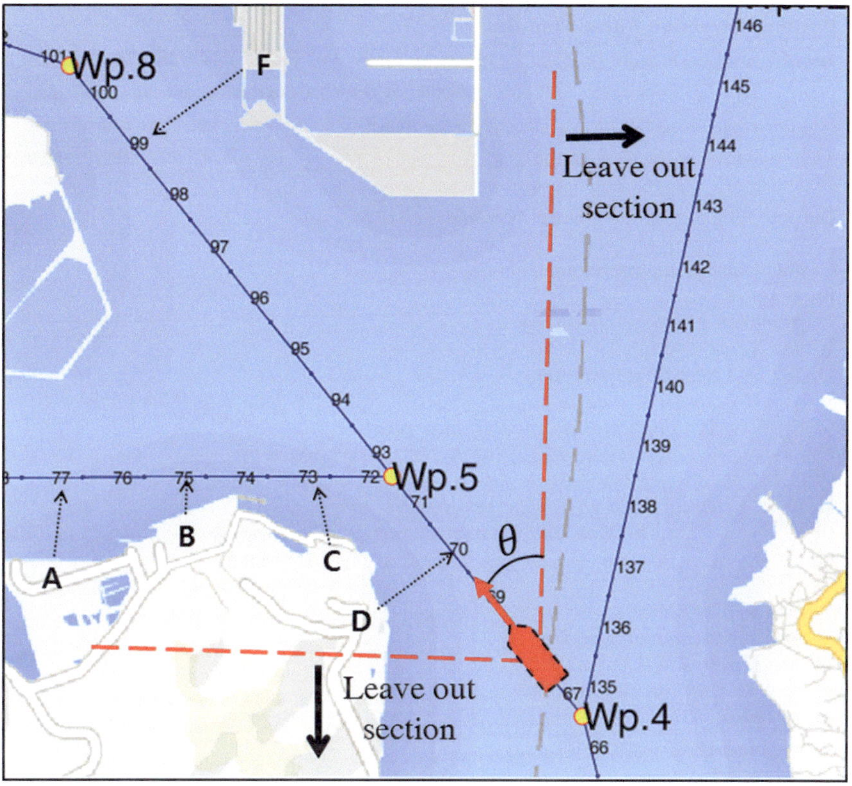

Fig. 3 The deep neural network for the candidate harbor proposal module

For performance comparison of the proposed method with the existing method, we choose the baseline method to choose the top frequent piers of the target vessels. When the baseline method was applied to the dataset, we obtained the following performance: 61.0% for the top 1 destination, 67.8% for the top 2 destinations, 72.4% for the top 3 destinations, and 76.2% for the top 4 destinations, and the accuracy of the 78.9% for the top 5 destinations.

When the proposed method was applied to the dataset, we obtained the following accuracies: 73.5% in the top 1 destination, 84.2% for the top 2 destinations, 87.2% for the top 3 destinations, and 89.0% for the top 4 destinations, the accuracy of the 89.8% for the top 5 destinations.

Figure 5 shows the accuracy over the top-ranked destinations for the proposed method and the baseline method obtained in the experiments. We observed that the proposed model gains higher accuracy by 10–15% than the baseline model.

Position-Direction Filter Module

Input : Ship destination probability vector by DNN *dest_dnn,* Latitude *ship_lat*, Longitude *ship_lon*, Course *ship_co*, Course deviation *co_dev,* Gate_Information structure data(latitude range, longitude range, inbound course range, inbound gates, outbound course range, outbound gates, adjacent pier) *gate_info.{lat_range, lon_range, inco_range, ingate, outco_range, outgate, adjpier}*

Output : Ship destination probability *output_dest*

```
// extract piers in ship course direction
tg ← Ø
gate_pier ← Ø
pier_vec ← Ø
For i = 1 to end[gate_info] do
    If ship_lat ∈ gate_info[i].lat_range and ship_lon ∈ gate_info[i].lon_range
        If ship_co ∈ gate_info[i].inco_range
            tg ← gate_info[i].ingate // get inbound gateline list
            gate_pier ← gate_info[tg].adjpier // get adjacent pier list
        elseif ship_co ∈ gate_info[i].outco_range
            tg ← gate_info[i].outgate // get outbound gateline list
            gate_pier ← gate_info[tg].adjpier // get adjacent pier list

// create pier list vector
For i = 1 to end[dest_dnn] Do
    If dest_dnn[i] ∈ gate_pier
        pier_vec ← true
    else
        pier_vec ← false

// filter dest_dnn into pier_vec
For i = 1 to end[pier_vec]
    If pier_vec [i]
        output_dest[i] ← dest_dnn[i]
    else
        output_dest[i] ← 0
```

Fig. 4 The pseudocode of position–direction filter module

5 Conclusions

In harbor water area, the prediction of ships' destination is important to estimate ship movements and collision risk. This paper presented a new prediction method that uses deep learning and real-time contextual information of ships. The proposed method uses a large volume of data from AIS and PORT-MIS systems to improve the accuracy of the prediction model.

The proposed method has shown that its performance is higher than the baseline method by 10–15%. The prediction of ships' destination lessons the burden VTS operators and ship navigators in estimating the future movements of ships and collision risks. We expect that the performance of the prediction model could be improved

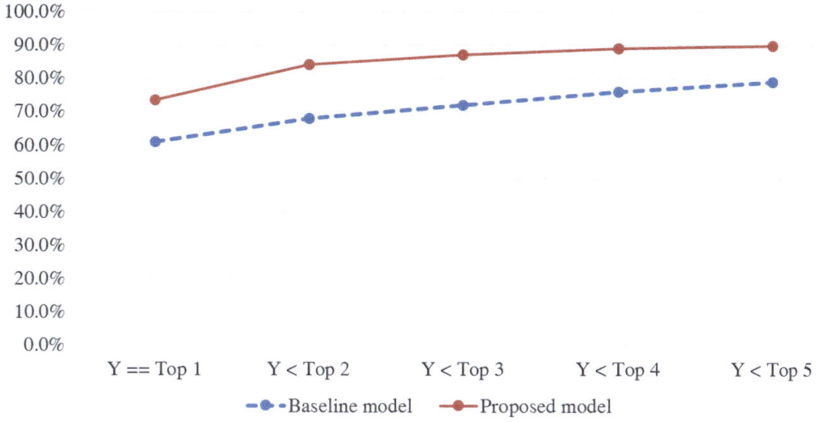

Fig. 5 Comparison of the proposed model and baseline model

by elaborating the deep neural network with architecture restructuring and further input variable articulations.

Acknowledgements This research was supported by Next-Generation Information Computing Development Program through the National Research Foundation of Korea (Grant no.: NRF-2017M3C4A7069432) and by Basic Science Research Programs through the National Research Foundation of Korea funded by the Ministry of Education (NRF-2016R1A6A3A11935806 and NRF-2015R1D1A1A01061062).

References

1. K. I. Kim, J. S. Jeong, and B. G. Lee: Study on the Analysis of Near-Miss Ship Collisions Using Logistic Regression. Journal of Advanced Computational Intelligence and Intelligent Informatics, 21(3), 467–473 (2017)
2. J. S. Jeong, G. K. Park, and J. S. Kim: Prediction Table to Improve Maritime Situation Awareness by VTS Opcrator. Proceedings of the International Conference on Fuzzy Theory and Its Applications (iFUZZY 2013). 479–481 (2013)
3. B. J. Rhodes, N. A. Bomberger, and M. Zandipour: Probabilistic associative learning of vessel motion patterns at multiple spatial scales for maritime situation awareness. Proceedings of the Information Fusion Conference, Qu'ebec, Canada (2007)
4. Pallotta G., Michele V. and Karna B.: Vessel Pattern Knowledge Discovery from AIS Data: A Framework for Anomaly Detection and Route Prediction. Entropy 2013, 15 (1), 2218–2245 (2013)
5. Ristic B., La Scala B., M. Morelande, N. Gordon: Statistical analysis of motion patterns in AIS Data: Anomaly detection and motion prediction. Proceedings of the 11th International Conference on Information Fusion, 40–46 (2008)
6. K. I. Kim, K. M. Lee, J. S. Jeong, and J. S. Kim: A Study on the Development of Decision Support System for VTS Operator using AIS data. Proceedings of the Marne Safety and Environmental Protection Conference, 239–239 (2016)

7. K. I. Kim and K. M. Lee: Ship Encounter Risk Evaluation for Coastal Areas with Holistic Maritime Traffic Data Analysis. Advanced Science Letters, 23(10), 9565–9569 (2017)
8. Yeosu Regional Ocean & Fisheries Administration's website: yeosu.mof.go.kr, 2015.

Design and Implementation of a Sunshine Duration Calculation System with Massively Parallel Processing

Woosuk Shin and Nakhoon Baek

Abstract The sun is one of the most important environmental elements for the human beings. To use the solar energy and sunlight, the sunshine duration is important, especially for architectural design and its related legal issues. To analyze the sunshine duration, there are several approaches, while they are not suitable to the modern computing environment. In this paper, we present a systematic way of calculating the sunshine duration based on the exact simulation of the sun paths. Furthermore, based on the modern massively parallel computer architectures and computer graphics techniques, we show the sunshine duration and we can generate the shadow accumulation maps in pseudo real time.

Keywords Sunshine duration · Massively parallel implementation · Parallel calculation

1 Introduction

The sun is actually one of the most important factors in human life. It provides heat and light directly, and grows plants and animals indirectly [1, 2]. It also enables to synthesize the Vitamin D for most vertebrates [3]. From the architectural point of

W. Shin · N. Baek (✉)
School of Computer Science and Engineering, Kyungpook National University, Daegu 41566, Republic of Korea
e-mail: oceancru@gmail.com

W. Shin
e-mail: mell03@naver.com

N. Baek
Software Technology Research Center, Kyungpook National University, Daegu 41566, Republic of Korea

N. Baek
Dassimey.com Inc., Daegu 41566, Republic of Korea

W. Lee and C. K. Leung (eds.), *Big Data Applications and Services 2017*,
Advances in Intelligent Systems and Computing 770,
https://doi.org/10.1007/978-981-13-0695-2_11

view, the sunshine from the sun is one of the most important elements to evaluate the values of real estates, as the form of sunshine duration [4].

Sunshine duration means the duration of sunshine in a given period for a given location on the Earth. It is actually a climatological indicator, and typically expressed as an averaged value over a long time period [5].

To measure the sunshine duration for a specific location, there are a set of historical approaches. In this paper, we present a sunshine duration framework, based on the massively parallel architecture and 3D graphics rendering features. Details of our system are presented in the following sections.

2 Previous Works

Most of previous works on the sunshine duration achieved with ray-tracing techniques [6]. Notice that these works are applicable only for a specific location. Additionally, these ray-tracing techniques are naturally applicable for a certain time, not for a time period.

In commercial markets, the "Ecotect Analysis" program [7], from the Autodesk Inc., had provided the sunshine duration analysis based on the ray-tracing techniques. Unfortunately, this module had run out of service since the year of 2015.

Other classes of the sunshine duration analysis approaches are based on the shadow generation algorithms from the computer graphics area. The shadows of the architectural buildings are generated and those shadows are analyzed with respect to the given time. Thus, these techniques are again not applicable to the time periods rather than a certain time.

Our goal is to develop more general and more efficient sunshine duration analysis system. And, this system may handle the whole time periods, such as the whole month of September, or the whole year. The following sections will show the design and implementation of our prototype system.

3 Design of the Sunshine Duration Calculation System

As shown in the previous sections, the previous approaches can calculate the sunshine durations, only for the user-specified time periods, and/or only for the specified location. In contrast, we aim to design a sunshine duration analysis system, to show the dynamically changing sunshine durations for any given time periods, with respect to all the surrounding architectural buildings.

3.1 Sun Path Simulation

Prior to other related calculations, we need to get the exact sun path, on the particular date, at the certain location. At the earth surface level, we can represent a specific sun direction as the azimuth angle and the elevation angle, as shown in Fig. 1.

We developed the exact sun path calculation routine, for any inputs of the surface positions (with the latitude and longitude values) on the earth, and the specific date and time.

3.2 Shadow Generation

Since a specific sun direction is calculated, we can simulate the shadows generated by the sun. In previous researches, they typically used some geometric calculations to get the shadows. In contrast, we will use the modern graphics hardware of the OpenGL depth buffer [8]. Current OpenGL SL (shading language) [9] features allow that we can precisely implement the shadow calculation algorithms as the OpenGL SL programs. Those OpenGL SL programs can be executed on the modern graphics processing unit (GPU) architectures, to achieve the massively parallel processing of the shadow generation.

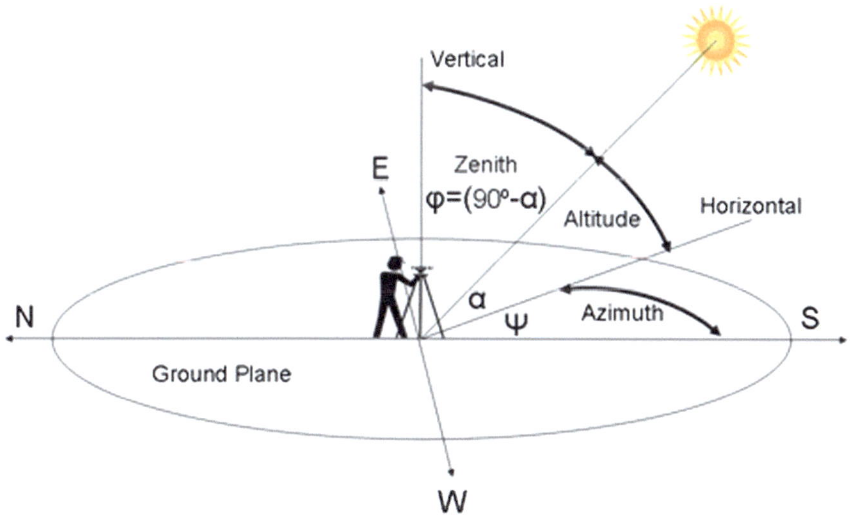

Fig. 1 Azimuth and altitude angles

3.3 Shadow Accumulation Maps

For the final sunshine analysis, we first need to record the shadows on the earth surface and also to accumulate the shadows for a given time period. For example, at the specific location, we can accumulate all the shadows generated by the surrounding architectural buildings. Analyzing the accumulated shadow, we can expect to calculate the sunshine duration. If there is no shadow for a specific location, it means there is no shadow at all, for the given time period. If there is a total shadow for a specific location, it shows that that location is under the shadow for the whole time, and it will be the worst case of the sunshine duration analysis.

Fortunately, the OpenGL SL can be used again to implement these shadow accumulation algorithms. Further, OpenGL features of the render-buffer and the frame-buffer provide the natural data structure for the accumulation of the graphics output including our shadows. In our case, we used a texture image as the render-buffer, and the whole shadows are accumulated on that texture image [10].

3.4 Sunshine Duration Analysis

Based on the accumulated shadow maps, we can provide more generalized query operations for the sunshine duration. As an advanced user experience, the user can click a specific part of an existing architectural building. Then, our system can project the 3D location on the accumulated shadow maps. Since the accumulated shadow map represents the shadow as the black color, and the sun-shined area as the white color, we can guess the ratio of sunshine duration, from the intensity of the projected location on the accumulated shadow maps. For a specific date, we can get the exact time period from the sunrise to the sunset. Multiplying the ratio value from the accumulated shadow map, we get the estimated sunshine duration for the specific date.

We started from the design of each module, which are explained in the previous subsections, we integrated all the modules into a system for sun-shine duration calculation. The experimental results are presented in the following section.

4 Implementation Results

Based on the designs presented in the previous sections, we have implemented the shadow generation and shadow map accumulation systems. The detailed experimental results are shown in this section.

4.1 Solar Motion Simulation

Our sun path simulation routine can generate the exact orbit trails of the sun. It also simulates the variations of the daylight durations, according to the changes in the solar motions. Figure 2 shows our simulation results for the different dates of March 1, June 1, September 1, and December 1, at a specific location in the northern temperate climate regions.

4.2 Calculating the Shadows

As the next step, we calculate the exact shadows of the architectures and objects, for the specific solar direction, as shown in Fig. 3.

For a specific date, we can simulate the whole solar orbit trail and calculate the shadows along that orbit trail. We can accumulate all the shadows, to finally get the accumulated shadow map of one day, as shown in Fig. 4a. We can further accumulate the shadow maps, to get the shadow maps for a specific month or for a whole year, as shown in Fig. 4b, c, respectively.

With the whole year shadow map, we can exactly evaluate the sunshine duration for a specific position. For example, the red point in Fig. 4c has the value of 17%

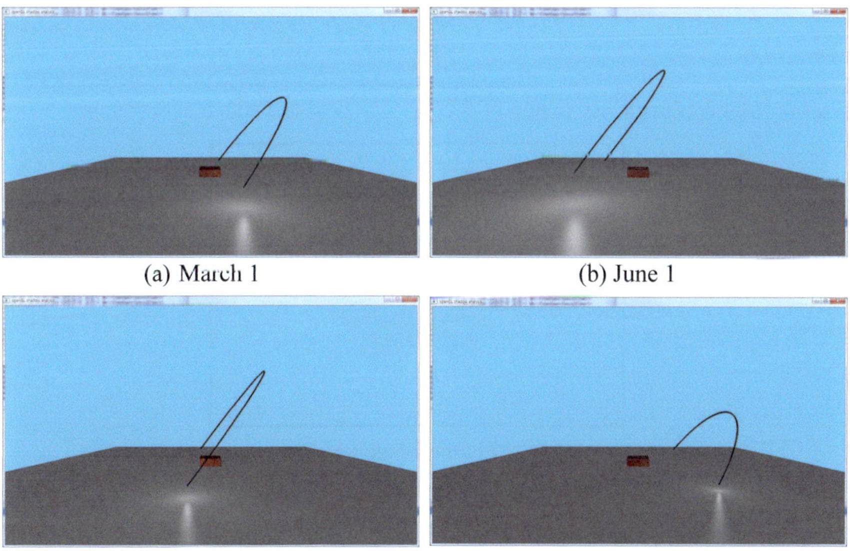

(a) March 1 (b) June 1

(c) September 1 (d) December 1

Fig. 2 Simulation results for sun paths on specific dates

Fig. 3 Shadow generation with respect to the specific solar position

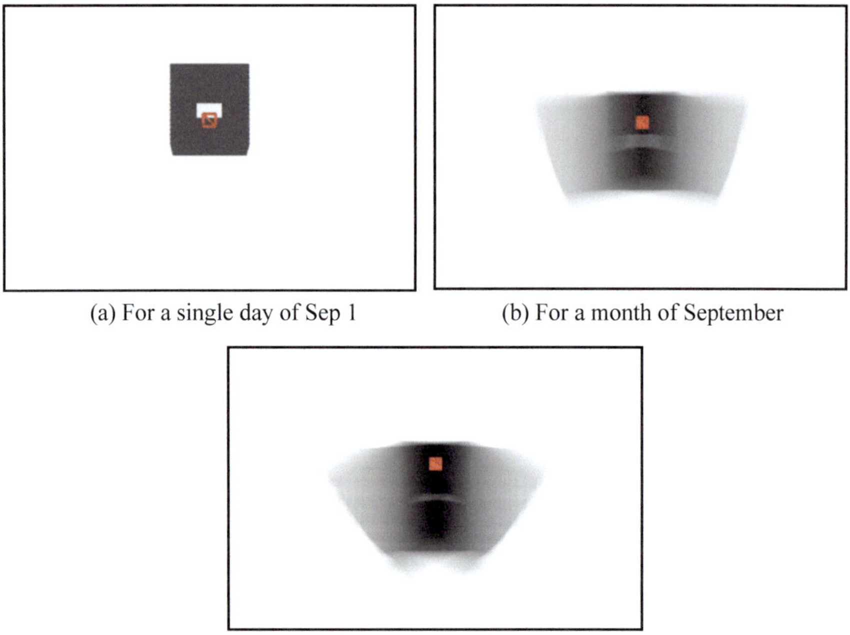

(a) For a single day of Sep 1

(b) For a month of September

(c) For the whole year

Fig. 4 The accumulated shadow maps

brightness. It is too short sunshine duration, and we can conclude that this area is not acceptable for human living areas.

5 Conclusion

Most of previous researches on the sunshine duration have focused on the sunshine duration analysis for a specific location. However, their result shows the limitations of these techniques. In this work, we present another sunshine duration analysis method, based on the sun path simulations and shadow accumulations.

Our method can fully utilize the modern graphics hardware support, to get the dramatic speed-ups with the massively parallel processing on the GPU architectures including CUDA (Compute Unified Device Architecture) [10] and DirectX [11]. Our next step will be in integrating more general terrains and more general architectural buildings for the shadow calculation. These expansions will make our system more practical, to be used with the up-to-date Geographic Information Systems (GIS) and other big-data processing systems.

Acknowledgements This research was supported by Basic Science Research Program through the National Research Foundation of Korea (NRF) funded by the Ministry of Education, Science and Technology (Grant No. 2016R1D1A3B03935488).

References

1. SPE: Global Market Outlook for Solar Power 2016–2020 (PDF). www.solarpowereurope.org. Solar Power Europe (SPE), (Retrieved 11 January 2016).
2. Duffy, J. F., Czeisler, C. A.: Effect of light on human circadian physiology. Sleep medicine clinics, 4.2: 165–177, (2009).
3. Holick, M. F.: Sunlight and vitamin D for bone health and prevention of autoimmune diseases, cancers, and cardiovascular disease. The American journal of clinical nutrition 80.6: 1678S–1688S, (2004).
4. Li, D. H. W., Wong, S. L.: Daylighting and energy implications due to shading effects from nearby buildings. Applied Energy, 84.12: 1199–1209, (2007).
5. World Meteorological Organization, Measurement of Sunshine Duration, Guide to Meteorological Instruments and Methods of Observation (PDF), WMO, (2008).
6. Li, DHW, Lau CCS, Lam JC: Predicting daylight illuminance by computer simulation techniques, Lighting Res. Technol., 36(2):113–129, (2004).
7. http://ecotect.com/ (retrieved in Oct 2017).
8. Segal, M., Akeley, K.: The OpenGL Graphics System: A Specification, Version 4.5 (Core Profile). Khronos Group, (2016).
9. Kessenich, J.: The OpenGL Shading Language, Version 4.40, Khronos Group, (2014).
10. NVIDIA, CUDA Runtime API, NVIDIA, (2017).
11. Luna F.: Introduction to 3D Game Programming with DirectX 12, Mercury Learning & Information (2016).

Developing 3D Annotation Features for 3D Digital Textbooks

Jin-Uk Jeong, Gae-A Ryu, Mihye Kim and Kwan-Hee Yoo

Abstract Digital textbooks (DTs) have been established as the major media for future and smart education in South Korea. In general, DTs are implemented using two-dimensional (2D) web-based platforms with embedded 3D content, including images, motion graphics, animations, and video and audio. Recently, these 2D-based DTs have evolved into a 3D-based interface by adopting various types of 3D virtual environments. Accordingly, a number of 3D DTs have been developed; however, these are mainly focused on the implementation of the basic features of DTs, including displaying text and images, viewing pages, zooming in and out, indexing, moving to a certain page, and finding a specific text or object. Furthermore, these have not yet been comprehensively implemented, and further development is required to provide more diverse input and annotation features to facilitate better dynamic interaction between students and DTs. Here, we introduce 3D annotation features that were designed and implemented to enable users to freely annotate on 3D DTs with various forms of input, facilitating dynamic user interactions. These include stylus writing, underlining, highlighting, drawing lines and shapes, and the insertion of textboxes, sound and video files, and voice memos. Students may highlight text or place notes to give commentary on the content of DTs using stylus writings, lines, shapes, text, multimedia files, and voice memos to aid their studying. The 3D annotation features were developed using the eXtensible 3D (X3D) standard, which is an XML-based international standard file format for representing scenes and 3D objects in computer

J.-U. Jeong · M. Kim
Department of Computer Science Education, Catholic University of Daegu, Gyeongsansi, Gyeongbuk, South Korea
e-mail: usmlike@nate.com

M. Kim
e-mail: mihyekim@cu.ac.kr

G.-A. Ryu · K.-H. Yoo (✉)
Department of Digital Informatics and Convergence, Chungbuk National University, 410 Seongbongro, Heungukgu, Cheongju, Chungbuk, South Korea
e-mail: khyoo@chungbuk.ac.kr

G.-A. Ryu
e-mail: reflexive@gmail.com

W. Lee and C. K. Leung (eds.), *Big Data Applications and Services 2017*, Advances in Intelligent Systems and Computing 770,
https://doi.org/10.1007/978-981-13-0695-2_12

graphics, extended from the virtual reality modeling language (VRML), along with the Java Applet and JavaScript. We aim to enhance student engagement in the learning process by supporting various forms of dynamic interaction features via a natural user interface. We believe that the approaches we describe provide a viable solution to enable 3D annotation on DTs.

Keywords Digital textbook · 3D digital textbook · X3D-based 3D digital textbook · 3D annotation

1 Introduction

With the continuing development of information technology (IT), including pervasive computing [1], cloud computing [2], and smart media devices [3, 4], educational systems are increasingly attracting innovative media for education and smart learning in the future, leading to an intelligent and adaptive learning system that will enhance learner efficiency and engagement [5]. Accordingly, in July 2011, the South Korean government established the Smart Educational Strategy Action Plan for a twenty-first century smart education system, and has launched a project to convert all existing printed textbooks (PTs) used in elementary, middle, and high schools into digital textbooks (DTs) suitable for smart learning environments by 2015 [5–8]. DTs are curriculum-based electronic textbooks that contain educational material used for teaching and learning in schools [9].

DTs are typically implemented with two-dimensional (2D) platforms and include all the functionality of PTs while providing a wider range of learning activities and environments by including various types of digital media features. With the adoption of 3D virtual environments in education systems, DTs are currently evolving to a 3D interface [10]. As a result, 3D DTs have been developed using the extensible 3D (X3D) standard by Hong et al. [10, 11]. The principle advantage of 3D DTs is the ability to realize a more realistic virtual environment, thus enabling users to perform various additional activities. For example, users can place virtual notes on a 3D object, as in temporarily attaching post-it notes. In addition, a 3D DT can present textbook content naturally without distortion of text or images because rotation, transition, and zoom operations can be freely applied with the 3D interface. Furthermore, dynamic interactions between users and 3D DTs are possible as both the content of DTs and objects embedded in the DTs exist in the same 3D space. Thus, a more realistic environment can be provided to students, enhancing their interest and engagement in the learning process [11]. The 3D DT developed by Hong et al. [10, 11] was for the subject of general computing in high schools in South Korea. The main features implemented in the 3D DT included displaying text and images, viewing pages, zooming in and out, indexing, moving to a particular page, finding specific text or objects, and embedding and playing 3D motion graphics, animations, and audio. The 3D DT was valuable in that the DT prepared guidance and directions for the development of a X3D-based DT. However, the 3D DT functionality demonstrated to date has focused on implementing the basic features of a DT in a 3D interface. Thus, further developments are required, particularly regarding annotations, to enable students to interact with DTs in a more fully interactive manner.

In this paper, we introduce 3D annotation features that were developed to allow users to freely annotate on 3D DTs via various forms of interaction. These include stylus writing, underlining, highlighting, drawing shapes, and the insertion of textboxes, multimedia files, and voice memos. Specially, head-mounted display (HMD) devices can be used to give interaction between users and the 3D DT. The content of this paper is based on the work described in [12] and the concept of the proposed approach was partially introduced in [13].

The remainder of this paper is organized as follows. In Sect. 2, we review the existing literature on 3D DTs and 3D annotation techniques. In Sect. 3, we present the basic architectures of the 3D DT for the process of 3D annotations, in addition to describing the implementation strategies of 3D annotations in relation to X3D nodes. In Sect. 4, we present 3D annotation features that were implemented, and Sect. 5 summarizes the paper and outlines directions for future work.

2 Related Works

2.1 3D DTs

3D content can enhance students' learning effectiveness by overcoming the limitations of the user experience in a 2D flat space, and also by facilitating learning engagement and motivation [11]. For these reasons, 3D content has been used in various ways in the field of education. In 2007, An and Kim developed a 3D web-based courseware for textbook content that enabled students to observe transportation routes used in the real-world and to experience the changes of civilizations using the Virtual Reality Modeling Language (VRML) and 3D Max [14]. In 2008, Yang et al. developed a science DT that shows the cut planes of multi-layer structures of 3D objects in real time as they are cut [15]. As mentioned in the previous section, Hong et al. also developed a 3D DT with various forms of 3D objects using X3D features for the subject of general computing [10, 11]. These 3D-based DTs can enhance students' motivation for learning by supporting dynamic user interaction with 3D objects [11]. However, these DTs focused on the implementation of the basic features of a DT in a 3D-based interface. Thus, further development is required, particularly for annotations to support more realistic user interactions.

Table 1 shows the 3D DT features implemented in the DT developed by Hong et al. [10, 11]. These are based on existing 2D DT features [7, 8]. The 3D DT features were implemented using the X3D functionalities.

Table 1 3D DT features implemented in the DT described in Refs. [10, 11]

Features	2D DT Features	3D DT features based on X3D
Display	Display texts and images	• Implement 3D DT texts and images in the same way as a PT using the Text Node and ImageTexture Node of X3D
	Single/double-page view	• Display DT content in single-page or double-page view using the settings feature of the ViewPoint Node
	Zoom in and out	• Using the mouse scroll wheel
	Fit a page to the screen	• Using the X3D viewpoint initialization key ("Esc" key)
	Fit a page to width/height	• Using the settings feature of the ViewPoint Node
	Page scroll	• Using the movement and viewpoint of the DT screen
	Indicate page thickness	• Indicate page thickness with an overlapped page view
	Display page number	• Show page thickness changes whenever a page is turned
	Text hiding	• By adjusting the transparency using the ScalarInterpolator Node
Input	Writing	• String input using the StringSensor Node
	Memo	• Using the JavaScript memo function
	View memo	• Display memos on a web browser
	Enter memo	• Input through a prompt dialog box
	Auto-save memo	• Save memos using web browser cookies
	Open memo	• Using the "get cookie" function of a web browser
	Delete memo	• Delete memos using web browser cookies
	Create a table of contents (TOC) for memos	• By saving in the form of a list of memos
	Adjust memo window	• By adjusting the size, move, and position of a memo window
Move	Move via TOC	• Move between Sections using the Anchor Node
	Move via the previous and next page buttons	• Page turning using the OrientationInterpolator Node • Page connection by the Inline Node • Movement button connection using the TouchSensor Node
	Page tuning	• Display the page turning motion
Search	Search content	• By linking to general search engines using the Anchor Node
	Sound effect	• Using the AudioClip Node

(continued)

Table 1 (continued)

Features	2D DT Features	3D DT features based on X3D
Multimedia	Multimedia	• Display pictures, voice, animations and playback of audio • Multimedia connection using the ImageTexture, AudioClip, MovieTexture, and Inline Nodes
	Interactive multimedia	• Playback of 3D multimedia data in the current window or a new window
Learning support	Hyperlinks	• Insert pages using the Inline Node • Connect hyperlinks using the Anchor Node
	Formative evaluation	• Hiding and revealing the answers to questions using the ScalarInterpolator Node

2.2 3D Annotation

3D virtual spaces are used in a number of fields, including games, politics, and society. Even an election campaign can be represented via a virtual reality service in Second Life. However, further development of 3D virtual technologies is also required for 3D object manipulation and collaboration in 3D spaces, as the number of applications of 3D spaces grows [16]. One of the major requirements for further development is that of 3D annotation techniques on 3D objects or 3D virtual spaces.

Annotation techniques in 3D virtual spaces have primarily been developed using 3D sketches to support collaboration between coworkers, especially in the field of design engineering. Bourguignon et al. developed an interactive sketch annotation system for 3D spaces for applications in education, design, architecture, and fashion, where 3D sketches were used to annotate 3D models [17]. Do and Gross developed a sketch-based interaction system for 3D domains using the Space Pen, which supports user annotation in the form of sketching written in Java 3D [18]. This system enables designers to mark or to add drawings on 3D models. Boujut and Dugdale developed a 3D annotation tool that supports collaborative design between designers by allowing them to leave comments on CAD models [16]. The tool was developed using Java 3D and SolidWorks, a CAD tool. Sookmyung Women's University has also performed studies on sketched-based annotation techniques for 3D object manipulation and cooperative work in 3D virtual spaces by resolving the recognition and phase difference of 3D objects [19]. Sin et al. improved sketch-based annotation techniques by exploring a sketch-box annotation technique suitable for 3D virtual spaces [20]. All of these reports focused on the development of more effective annotation techniques for 3D models using a sketch-based interface to support collaboration between designers in engineering fields. Hence, the annotation features of these sketch-based approaches were limited, and it is not straightforward to apply them to other fields.

Here, we report a different annotation approach for X3D-based DTs to allow students to freely annotate DTs using various types of input via X3D. Here, the term annotation can be defined as notes, memos, and drawings attached or inserted on the content of 3D DTs. For example, highlighted text or virtual post-it notes to provide a commentary on the content of DTs using lines, circles, texts, and voice with a view to facilitating study are included in the annotation features in our approach.

X3D is an XML-based international standard file format for representing 3D computer graphics objects in VRML [21, 22]. There exist a number of 3D standard file formats, including Universal 3D (U3D), JAVA 3D, and Coin3D. Among these, only X3D is an XML-based file format comparable with existing XML DT formats with a rich variety of functionalities [11]. Moreover, it is comparable with VRML 97 and supports a dedicated platform independent web browser (i.e., BS Contact) with capability for integration with external programs. For these reasons, we used X3D as the file format for 3D, which was also used for the work reports in Refs. [10, 11]. The X3D feature set includes 3D graphics, 2D graphics, animation, audio and video, user interaction, navigation, user-defined objects, scripting, and networking, in addition to physical simulation, geospatial positioning, CAD geometry, layering, support for programmable shaders, and particle systems [22]. A more detailed description of X3D can be found at the Web3D Consortium [21]. The X3D features are provided in the object hierarchy, with nodes grouped into components with the related objects [22].

3 Design of 3D Annotation Features for 3D DTs Using X3D

3.1 Main Features of 3D Annotations for X3D-Based 3D DTs

We determined the 3D annotation features for inclusion in DTs based on the input and multimedia functions of 2D-based DTs from the Korea Education and Research Information Service (KERIS) [7, 8]. We then designed the annotation features using X3D nodes on the X3D-based DT developed by Hong et al. [10, 11]. Table 2 shows the 3D annotation features we designed for the X3D-based DTs and the X3D nodes used for the implementation of each feature. The memo annotation feature was implemented with a memo pad [10, 11] and, in addition, we support a voice memo feature.

3.2 X3D Nodes for 3D Annotation Features

To realize 3D annotations in an X3D-based DT, we utilized the TouchSensor, Script, and Transform nodes of X3D and extended them appropriately for each annotation feature.

- *TouchSensor*. The interactions of X3D consist of sensing user actions and prompting appropriate responses. We used the TouchSensor node of X3D to recognize

Table 2 3D annotation features for 3D DTs and the corresponding X3D nodes

Features	2D DT features	3D nodes for each 3D annotation feature
Stylus writing	Writing on 3D DTs using a stylus pen or mouse	LineSet
Highlight	Highlighting using a highlight pen	TriangleStripSet
Textbox	Input strings using a keyboard	Text
Underline or line	Underlining using a stylus pen or mouse	LineSet
Shapes	Creating and drawing various shapes on 3D DTs	LineSet
Multimedia	Insert sound and video files on 3D DTs	MovieTexture
Voice memo	Insert voice memos (voice recording) using a microphone	Java Applet

Table 3 TouchSensor node used to obtain 3D coordinates

```
<Transform DEF="touchPlane">
    <Shape>
        <Box size="8 12 0.0015"/>
        <Appearance>
            <Material transparency="1"/>
            </Appearance>
    </Shape>
    <TouchSensor DEF="Input" enabled="false"/>
</Transform>
```

user action and movements on a 3D space using a mouse. To obtain 3D coordinate values for the annotations, we added a touch sensor (i.e., touch panel) over the 3D DT, and defined the touch sensor in the Transform node, as shown in Table 3. For keyboard events, we used the StringSensor node.

- *Script*. Application programmers can be used to define additional fields and features when instantiating the Script node within a scene in X3D. That is, programmers may define the interface and behavior of a Script node. In X3D, sensors detect various kinds of user interaction and produce events to ROUTE within a scene. Here, programmers define how the events describing user interaction are handled using a programming language that the X3D browser supports. We created Script nodes for each annotation feature, defining the interface and behavior of the Script nodes to handle incoming events using JavaScript.
- *Menu*. We implemented seven annotation features, as listed in Table 2, which were provided via a menu interface. Thus, users must select an annotation type from the given menu to make an annotation.

3.3 3D Annotation Process

Figure 1 shows the overview of our 3D annotation process for X3D-based DTs. First, a user should select the annotation type (icon) from the annotation menu. The system then initializes the TouchSensor node and the respective Script node corresponding to the annotation type, and produces events to ROUTE within a scene. When a user clicks on a starting position on the 3D DT using a mouse, TouchSensor reads the 3D coordinate values of the start position and executes the respective Script node. Then, the Script creates an X3D node corresponding to the selected annotation type. For example, when a user selects a stylus writing in the annotation menu, the Script will create a LineSet node, as described in Table 2. The Script then assigns these coordinate values to the property values of the respective X3D node in real time according to the movements of the mouse, drawing and displaying the annotation results on the 3D DT. Here, moving the mouse while the sensor is activated (i.e., mouse dragging motion) will cause continuous generation of output events. Then, the Script adds the created X3D node (e.g., LineSet, TriangleStripSet, Text, MovieTexture nodes) to the Transform node of X3D. The X3D nodes encoded in the Transform node for annotations will be viewed on the DT, where the annotation was inserted. That is, X3D will display the annotations on the DT, along with the origin contents.

When a keyboard event occurs, the StringSensor node is activated. To change a line thickness, we use the PlaneSensor node of X3D. The X3D nodes registered in the Transform node can be deleted and the annotation color and thickness for annotations may also be altered.

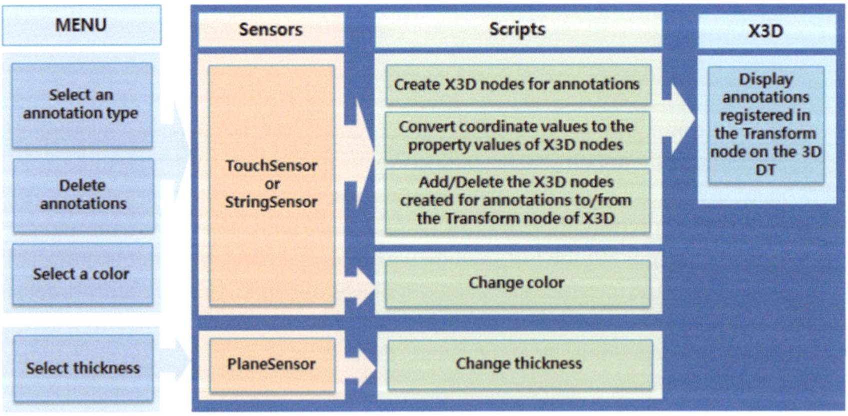

Fig. 1 Overall process of 3D annotations on an X3D-based DT

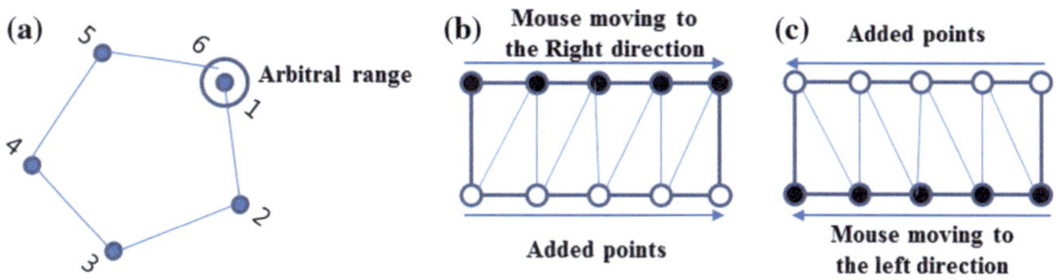

Fig. 2 **a** An example of drawing a shape, **b** highlighting in the right direction (**b**), and **c** highlighting in the left direction

3.4 Design for 3D Annotation Features

- *Stylus Writing, Underline, and Shapes*. We designed these annotations using the LineSet node of X3D. Thus, the internal processes of these annotations are similar. The differences lie in the control of mouse clicks. In the case of stylus writing, all the 3D coordinates followed by the mouse moving through a 3D space are continuously transmitted to the LineSet node created by the respective Script node, because all the data for the trajectory of the mouse are required. For an underline or a line, two 3D coordinate points, one from where the mouse was clicked for the first time and the other from where the mouse was released, will be transmitted to draw a line. In the case of shape drawing, whenever a vertex of a shape is clicked, the 3D coordinate values of the vertex will be transmitted to the corresponding LineSet node. When a vertex input from a mouse click is completed, a shape will be drawn by connecting the vertices. If a point is clicked within a predefined arbitrary range around the first clicked point, the point will be recognized as a signal of the final vertex input and will be identified with the first clicked vertex. For example, if the vertices of a shape are clicked from 1 to 5 in order, and then point 6 is clicked, the annotation system will identify point 6 with point 1 and will draw a pentagram by connecting points from 1 to 5 sequentially, as shown in Fig. 2a.
- *Highlight*. To represent lines, X3D uses vector formats that cannot control line thickness. Thus, we process a highlight from a rectangle using the TriangleStripSet node of X3D. The TriangleStripSet node creates a rectangle by connecting the three coordinate points that were taken most recently. However, the coordinate points followed by the mouse moving for a highlight cannot generate a rectangle. Accordingly, we use additional points to generate a rectangle in the direction that the mouse moves, as shown in Fig. 2b, c. The vertical extent of the rectangle will be determined by the thickness of the highlight.
- *Textbox*. We designed textbox input using the Text node of X3D. A textbox is created at a 3D point where the mouse was clicked, obtained using the TouchSensor node. Strings are then entered in the textbox using the StringSensor node that provides a string-based interface from the user's keyboard. Each key press is collected until the enter key is pressed, which is used to signal the end of the input.

- *Multimedia*. Users can annotate on the content of the DT by inserting audio or video files. The location where a multimedia file is inserted is obtained using the TouchSensor node (i.e., from a mouse click) and the file path from the StringSensor node. The multimedia file is inserted to the DT using the MovieTexture node of the texture component of X3D. To play multimedia files, the startTime and stopTime properties of the texture component are used.
- *Voice Memo*. To record a voice memo, hardware control is necessary to control a microphone and a sound card. However, X3D does not support hardware control features. For this reason, we developed a voice memo using the java.applet and linked it to the DT using the Anchor node of X3D. We implemented a voice recording feature using the AudioFormat and DataLine classes, and a playback feature using the AudioClip class in Java. Java applets are set to disable the hardware control because of security issues. Thus, using the keytool and jarsigner utilities included in the Java Development Kit, we convert the applet implemented for voice memos to a Signed applet that allows users to control hardware.

4 Implementation Results of 3D Annotation Features

We implemented the 3D annotation features listed in Table 2 using the 3D DT developed by Hong et al. [10, 11]. The annotation system was developed using a Windows 7 platform with the X3D browser (BS Contact) and X3D Edit for the X3D nodes, Java Applet, and JavaScript.

Figure 3 shows the screen configuration of the 3D DT, including the Table of Contents (TOC) on the left side of the screen, textbook contents in the double-page view, and the menu for 3D annotations. We added the annotation menu to the existing DT using the Inline node of X3D. The annotation menu at the bottom of the screen includes a palette for selecting colors, a scroll bar for selecting the thickness, an eraser for deleting annotations, a pen for stylus writing, and a highlighter pen for highlighting, as well as icons for textboxes, lines, shapes, and video and video file inputs, as indicated in the figure. A feature icon for voice memos was included, as shown in the top right of the screen, shown in Fig. 3.

- *Stylus Writing and Highlighting*. The most frequently used annotation features in a DT are expected to be stylus writing and highlighting. Figure 4a shows examples of stylus writing and Fig. 4b shows examples of highlighting. Table 4 lists the JavaScript source code used to obtain a block of an X3D node when the mouse is clicked. The received block is added to X3D using the Browser.createVrmFromString() function and the variable btnClick is set to true to maintain the status of mouse click. If the mouse button is released, then coordNo is set to 0. Table 5 lists a function that transfers the 3D coordinates to an X3D node for stylus writing.

Users may change the highlighting color and thickness. Table 6 shows the function used to generate 3D coordinates for highlighting representation. To represent

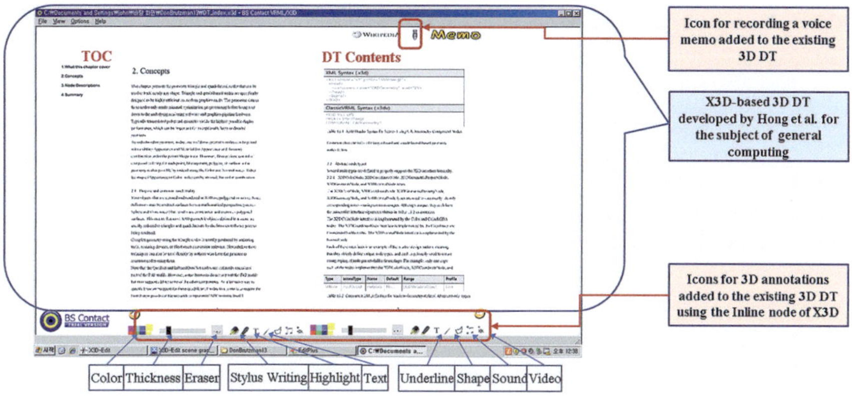

Fig. 3 Screen configuration of the 3D DT

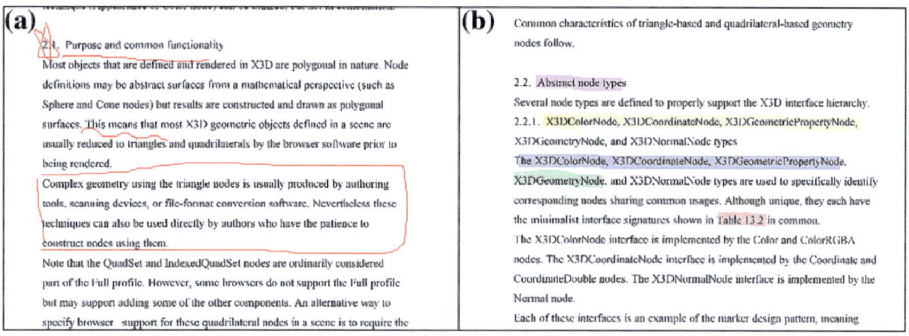

Fig. 4 Examples of **a** stylus writings and **b** highlighting

Table 4 JavaScript source code to handle mouse clicks

```
function isActive(value, timestamp) {
    if (value ==true) {
            AnnotationNode.addChildren = Browser.createVrmlFromString(getBLOCK());
            btnClick = true;
            nodeNo = nodeNo + 1;
    } else {
            btnClick = false;
            coordNo = 0;
    }
}
```

highlighting, the function obtains and generates 3D coordinates based on mouse movement in real-time, checking the color information, and transmits these data to the TriangleStripSet node. If the mouse is moved to the right, then right = true; otherwise, right = false.

Table 5 Function to transfer 3D coordinates to an X3D node for stylus writing

```
function freeLineCoordChange(value, timestamp) {
    if (btnClick == true) {
            AnnotationNode.children[nodeNo1].children[0].geometry.coord.point[coordNo]
                = new SFVec3f(value.x, value.y, value.z-0.0001);
            AnnotationNode.children[nodeNo-1].children[0].geometry.color.color[coordNo]
                = selectedColor;
            AnnotationNode.children[nodeNo-1].children[0].geometry.vertexCount[0] = coordNo + 1;
            coordNo = coordNo + 1;
    }
}
```

Table 6 The function used to generate 3D coordinates for the highlight representation

```
function fluCoordChange(value, timestamp) {
    if (btnClick == true) {
        AnnotationNode.children[nodeNo-1].children[0].geometry.coord.point[coordNo]
            = new SFVec3f(value.x, value.y, value.z-0.0001);
        AnnotationNode.children[nodeNo-1].children[0].geometry.color.color[coordNo] =
        selectedColor;

        if (AnnotationNode.children[nodeNo-1].children[0].geometry.coord.point[coordNo].x
            - AnnotationNode.children[nodeNo-1].children[0].geometry.coord.point[coordNo-2].x >
0) {
            right = true;
        }
        else if (AnnotationNode.children[nodeNo-1].children[0].geometry.coord.point[coordNo].x
            - AnnotationNode.children[nodeNo-1].children[0].geometry.coord.point[coordNo-2].x <
0) {
            right = false;
        }
        if (right == true){
            AnnotationNode.children[nodeNo-1].children[0].geometry.coord.point[coordNo+1]
                = new SFVec3f(value.x, value.y-thick, value.z-0.0001);
            AnnotationNode.children[nodeNo-1].children[0].geometry.color.color[coordNo+1]
                = selectedColor;
            AnnotationNode.children[nodeNo-1].children[0].geometry.stripCount[0] = coordNo + 2;
        }
        else {
            AnnotationNode.children[nodeNo-1].children[0].geometry.coord.point[coordNo+1]
                = new SFVec3f(value.x, value.y+thick, value.z-0.0001);
            AnnotationNode.children[nodeNo-1].children[0].geometry.color.color[coordNo+1]
                = selectedColor;
            AnnotationNode.children[nodeNo-1].children[0].geometry.stripCount[0] = coordNo + 2;
        }
        coordNo = coordNo + 2;
    }
}
```

- *Textbox*. Users may also enter text (i.e., strings) via the keyboard. If the textbox icon from the annotation menu is selected, a textbox will be created at the position where the mouse was clicked, which is determined using the TouchSensor node. Strings can then be entered in the textbox using the keyboard via the StringSensor node. Each character key press is collected until the enter key is pressed. Figure 5 shows examples of textbox inputs and Table 7 lists the functions implemented to handle the textbox input.

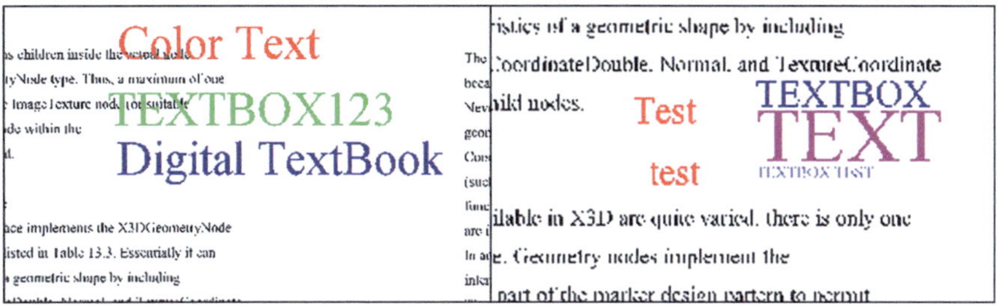

Fig. 5 Examples of textbox inputs with different colors and sizes

Table 7 The functions used to process a textbox input

```
function textIsActive(value, timestamp) {
    AnnotationNode.addChildren= Browser.createVrmlFromString(getBLOCK());
    nodeNo = nodeNo + 1;
}
function textBoxCoordChange(value, timestamp){
    BtnLoc = value;
}
function textBoxInput(value) {
    AnnotationNode.children[nodeNo-1].children[0].children[0].geometry.string = new
MFString(value);
}
```

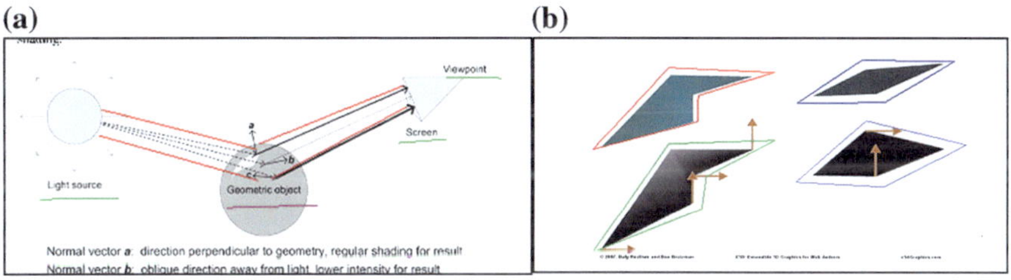

Fig. 6 Examples of **a** lines and **b** shapes drawn using a combination of stylus inputs and lines

- *Line and Shape*. Users can underline text where they wish to emphasize something, or may draw a line to connect two arbitrary points on the 3D DT. After selecting the line icon from the annotation menu, the user may click to define a starting point of a line using the mouse, and then drag the mouse until the endpoint of the line and release the mouse button. A line will then be drawn by connecting the starting and end points. The left screen of Fig. 6 shows examples of lines and Table 8 lists the functions used to implement line annotation using JavaScript.

Users may also draw a variety of shapes on the 3D DT. The right screen of Fig. 6 shows examples of shapes drawn using a combination of stylus and line input. Table 9 lists the function used to implement drawing shapes. When the mouse is clicked for the first time, the variable firstClick will be set to false and the first clicked point will be assigned to the variable StartPoint. If the assigned StartPoint exists within a

Table 8 The function used for drawing lines

```
function lineCoordChange(value, timestamp) {
    if (btnClick == true) {
        AnnotationNode.children[nodeNo-1].children[0].geometry.coord.point[coordNo]
            = new SFVec3f(value.x, value.y, value.z-0.0001);
        AnnotationNode.children[nodeNo-1].children[0]        .geometry.color.color[coordNo]        =
selectedColor;
        AnnotationNode.children[nodeNo-1].children[0].geometry.vertexCount[0] = coordNo + 1;
        coordNo = 1;
    }
}
```

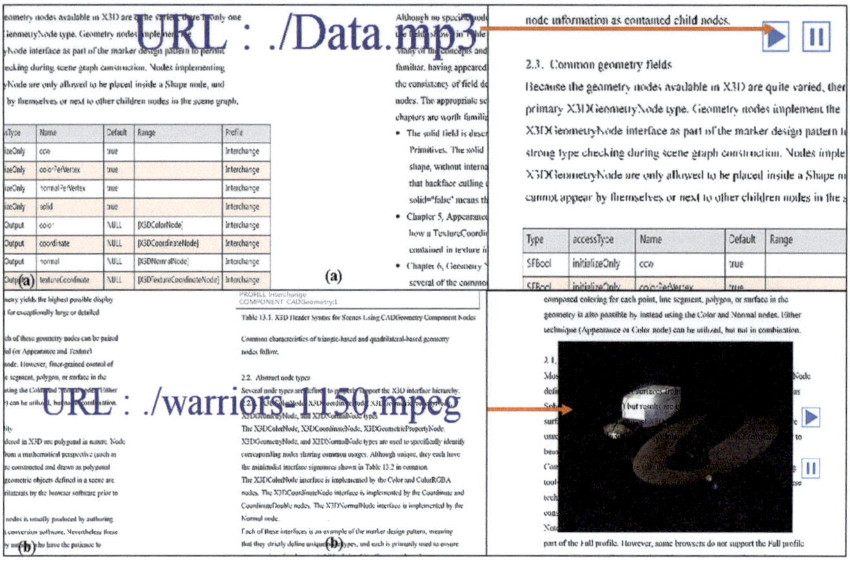

Fig. 7 Examples of **a** inserting a sound file and **b** inserting a video file

predefined arbitrary range, the point will be recognized as a signal for the final vertex input and will be identified with the first clicked point of the shape.

- *Audio and Video*. Users can also insert sound, music, or video files into the 3D DT. These features were implemented using the MovieTexture node of X3D. A user may click a point at which they want to place a multimedia file after selecting the sound or video icons from the annotation menu, and a URL input screen will appear on the DT. When the file path for the multimedia file is entered, the system will add play and pause buttons at the point where the mouse was clicked. In the case of a video file, the system will also add a video screen, as shown in the bottom right screen in Fig. 7. Tables 10 and 11 show the functions implemented to insert sound and video files into the DT, respectively.
- *Voice Memo*. X3D does not support hardware control features, such as a microphone and sound card, so we recorded voice memos using a Java applet and linked

Table 9 Function to process the shape annotation on the 3D DT

```
function polyCoordChange(value, timestamp) {
    if (btnClick == true) {
        AnnotationNode.children[nodeNo-1].children[0].geometry.coord.point[coordNo]
            = new SFVec3f(value.x, value.y, value.z-0.0001);
        AnnotationNode.children[nodeNo-1].children[0].geometry.color.color[coordNo] =
selectedColor;
        AnnotationNode.children[nodeNo-1].children[0].geometry.vertexCount[0] = coordNo + 1;
        if (firstClick == true) {
            startPoint = value;
            firstClick = false;
        }
        else {
            If ((startPoint.x+rad > value.x && startPoint.x-rad < value.x) && (startPoint.y+rad >
value.y
              &&startPoint.y-rad<value.y)&&(startPoint.z+rad>value.z && startPoint.z-rad<value.z))
{
                AnnotationNode.children[nodeNo-1].children[0].geometry.coord.point[coordNo]
                    = startPoint;
                AnnotationNode.children[nodeNo-1].children[0].geometry.color.color[coordNo]
                    = selectedColor;
                AnnotationNode.children[nodeNo-1].children[0].geometry.vertexCount[0] = coordNo
+ 1;
                firstClick = true;
                coordNo = 0;
            }
        }
        btnClick = false;
    }
    else {
        if(firstClick == false) {
            AnnotationNode.children[nodeNo-1].children[0].geometry.coord.point[coordNo+1]
                = new SFVec3f(value.x, value.y, value.z-0.0001);
            AnnotationNode.children[nodeNo-1].children[0].geometry.color.color[coordNo+1]
                = selectedColor;
            AnnotationNode.children[nodeNo-1].children[0].geometry.vertexCount[0] = coordNo + 2;
        }
    }
}
```

them to the 3D DT using the Anchor node of X3D. Figure 8 shows screen captures for recording a voice memo. If the icon for the voice memo shown in the top right of the DT screen is selected, an applet window will open, as shown in Fig. 8. First, the user should enter a file name in the bottom of the window to save the audio recording, and then start recording the voice memo by pressing the Capture button. During recording, the Stop button will become enabled, while the Capture and Play buttons become disabled. Pressing the Stop button terminates the recording, and the voice memo will then be saved using the file name that was supplied earlier, and will be displayed in the memo list shown in the middle-left of the window. A voice memo can be played using the Play button after selecting it from the memo list.

Table 10 Function to insert sound files to the 3D DT

```
function getMUSIC() {
   var addfile = 'Transform {
     children [
       Transform {
         translation 6.9 6.8 –0.065 children
           [ DEF Moviestart'+MnodeNo+' Shape    {
             geometry Box { size 0 0 0 }
             appearance Appearance {
               texture DEF Movie'+MnodeNo+' MovieTexture {
                 repeatS FALSE repeatT FALSE url[" '+fileLoc+' "] } } } ] }
       Transform {
         translation'+BtnLoc.x+' '+BtnLoc.y+' '+(BtnLoc.z+0.1)+'
         children [Shape {
           geometry Box { size 0.4 0.4 0 }
           appearance Appearance {
             texture ImageTexture {
               url ["../images/btn_start.jpg"] } } }
             DEF MovieStart'+MnodeNo+' TouchSensor {} ] }
       Transform {
         translation '+(BtnLoc.x+0.6)+' '+BtnLoc.y +' '+(BtnLoc.z+0.1)+'
         children [ Shape {
           geometry Box { size 0.4 0.4 0 }
           appearance Appearance {
             texture DEF btn_stop'+MnodeNo+'ImageTexture {
               url ["../images/btn_stop.jpg"] } } }
             DEF MovieStop'+MnodeNo+' TouchSensor {}] }
     ]
     ROUTE MovieStart'+MnodeNo+'.touchTime TO Movie'+MnodeNo+'.startTime
     ROUTE MovieStop'+MnodeNo+'.touchTime TO Movie'+MnodeNo+'.stopTime
   }';
   return addfile;
}
function mediaCoordChange(value, timestamp) {
   BtnLoc = value;
}
function keyboard(value) {
   if (!value.equals("\n")) {
           KeyString[0] = 'URL :' + value;
           fileLoc = value;
   }
}
function multimediaIsActive(value,timestamp){
   if(value == true){
           StrEnable = true;
           KeyString[0] = 'URL : ';
   }
}
```

Table 11 The function used to insert video files to the 3D DT

```
function getMOVIE() {
    var addfile
      = 'Transform {
        children [
          Transform {
            translation '+(BtnLoc.x-2.9)+' ' +(BtnLoc.y-0.4)+' '+BtnLoc.z+'
            children [ DEF Moviestart'+MnodeNo+'
              Shape {
                geometry Box { size 5 3.75 0 }
                appearance Appearance {
                  texture DEF Movie'+MnodeNo+'
                  MovieTexture {
                    repeatS FALSE repeatT FALSE url[" '+fileLoc+' " ] } } } ] }
          Transform {
            translation '+BtnLoc.x+' '+BtnLoc.y+' '+(BtnLoc.z+0.1)+'
            children [
              Shape {
                geometry Box { size 0.4 0.4 0 }
                appearance Appearance {
                  texture ImageTexture {url ["../images/btn_start.jpg"]
              } } }
              DEF MovieStart'+MnodeNo+' TouchSensor { }
          ] }
          Transform {
            translation '+BtnLoc.x+' '+(BtnLoc.y-1)+''+(BtnLoc.z+0.1)+'
            children [
              Shape {
                geometry Box { size 0.4 0.4 0 }
                appearance Appearance {
                  texture ImageTexture {url ["../images/btn_stop.jpg"]
              } } }
              DEF MovieStop'+MnodeNo+' TouchSensor { }
        ] } ]
        ROUTE MovieStart'+MnodeNo+'.touchTime TO Movie'+MnodeNo+'.startTime
        ROUTE MovieStop'+MnodeNo+'.touchTime TO Movie'+MnodeNo+'.stopTime
      } ] }';
      return addfile;
}
```

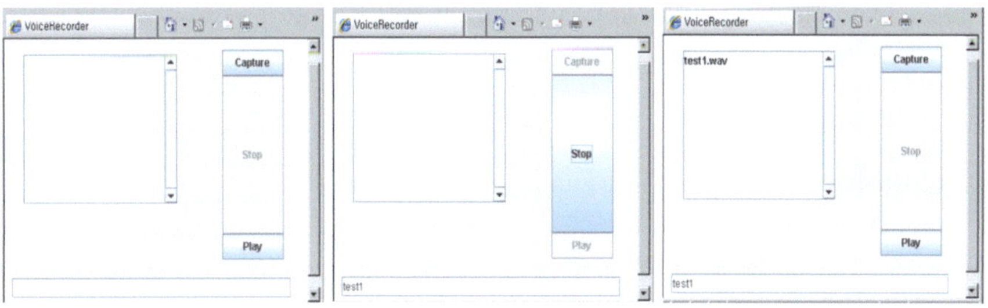

Fig. 8 An example recording a voice memo

Table 12 The functions used to process voice memos

```
private void captureAudio() { // voice recordings
    try{
        audioFormat = getAudioFormat();
        DataLine.Info dataLineInfo =   new DataLine.Info(TargetDataLine.class, audioFormat);
        targetDataLine = (TargetDataLine)AudioSystem.getLine(dataLineInfo);
        new CaptureThread().start();
    }catch (Exception e) {
        e.printStackTrace();
        System.exit(0);
    }
}
private AudioFormat getAudioFormat() { // acquiring signals form a microphone
    float sampleRate = 8000.0F;
    int sampleSizeInBits = 16;
    int channels = 1;
    boolean signed = true;
    boolean bigEndian = false;
    return new AudioFormat(sampleRate, sampleSizeInBits, channels, signed, bigEndian);
}
private void playAudio() { // playing a voice memo
    try{
        File waveFile = new File(SelectedFile);
        ac = Applet.newAudioClip(waveFile.toURL());
        ac.loop();
   }catch(Exception e) {
        e.printStackTrace();
   }
}
```

Table 12 lists the functions associated with the voice memo annotation. The captureAudio() function records audio by acquiring signals from a microphone using the getAudioFormat() function and saves them into the given file. The playAudio() function is for playing a saved voice memo.

Figure 9 shows examples of 3D objects in the DT annotated using stylus writing and lines, viewed from different angles. Such annotated features can provide a more realistic learning environment for students, enhancing their interest and motivation in the learning process.

5 Conclusion

The principle advantage of 3D annotations is the ability to realize a more realistic virtual education environment, thus enabling students to perform various additional learning activities on 3D DTs. We have described the design and implementation of 3D annotations on X3D-based DTs. The objective was to develop 3D annotation features that can enable students to freely place notes into 3D DTs via various forms of interactions. The aim was to enhance student interest and engagement in the learning process by supporting a more realistic environment and various forms of

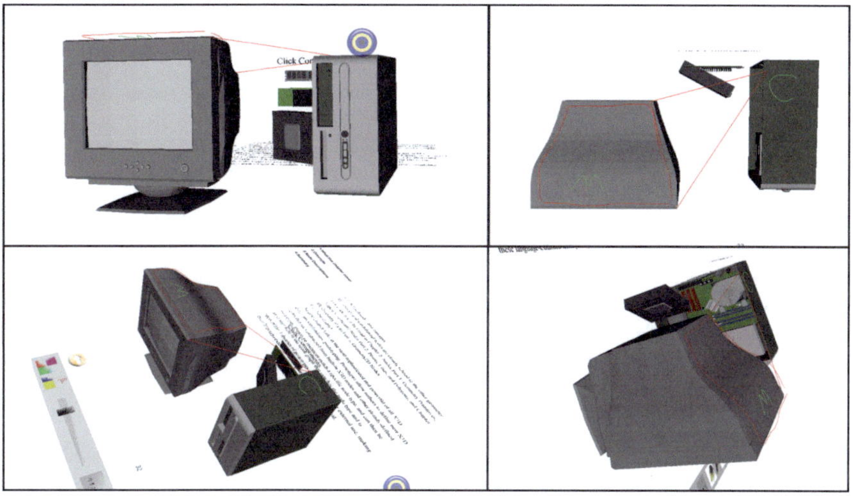

Fig. 9 3D annotation examples on the 3D DT

dynamic interaction between users and DTs. The developed 3D annotation features include stylus writing, highlighting, underlining, drawing shapes, and the insertion of textboxes, multimedia files, and voice memos. Students can freely highlight text or place notes to give commentary on the content of DTs using text, stylus writings, lines, circles, shapes, multimedia files, or voice to aid their studying.

Although the development of these 3D annotations are valuable and may provide some guidance and directions for 3D interactions between users, there are a number of fundamental issues that require further development. First, X3D cannot control hardware because it is itself configured as a web-based 3D environment. As a result, X3D is not able to save annotation data directly to disk. To handle this issue, an additional interface is required, which has issues for portability. Second, in X3D, the perception of an object has a higher priority than the recognition of a sensor. Thus, when a node is generated for an annotation and overlaps with a sensor, abnormalities may occur in displaying the contents. Third, there exist issues in handling 2D-based input devices, such as the mouse and stylus writing pen, whereas 3D annotation is made in a 3D space. Thus, the development of input devices suitable for 3D environments is desirable. However, this study is a significant step towards the development of 3D annotation techniques for 3D, providing an intuitive user interface, which may facilitate an enhanced smart learning environment.

Acknowledgements This research was financially supported by the ICT R&D program of MSIP/IITP [2016000121002100l, Standard Development of HMD Based VR Service Framework] and by the "Standard Technology Improvement Program" through the Ministry of Trade, Industry and Energy (No. 10053638).

References

1. J. Luo, F. Dong, J. Cao, A. Song, "A context-aware personalized resource recommendation for pervasive learning," Cluster Comput., Vol. 13, No. 2, pp. 213–239, 2010.
2. J.-M. Kim, H.-Y. Jeong, I. Cho, S. M. Kang, J. H. Park, "A secure smart-work service model based OpenStack for Cloud computing," Cluster Comput. https://doi.org/10.1007/s10586-013-0251-1, 2013.
3. S. Choi, K. Chung, H. Yu, "Fault tolerance and QoS scheduling using CAN in mobile social cloud computing," Cluster Comput. https://doi.org/10.1007/s10586-013-0286-3, 2013.
4. D. Lee, H. Lee, D. Park, Y. Jeong, "Proxy based seamless connection management method in mobile cloud computing," Cluster Comput. https://doi.org/10.1007/s10586-013-0249-8, 2013.
5. Ministry of Science, Education and Technology: Road to an Advanced Country: Smart Educational Strategy Action Plan. Republic of Korea, 2011. http://www.mest.go.kr/web/1127/ko/board/view.do?bbsId = 192&mode = view&boardSeq = 25734.
6. Korean Education & Research Information Service: Digital Textbook, http://www.dtbook.kr/renew/english/index.htm (accessed on November 2013).
7. H. Byun, K. Yoo, J. Yoo, J. Choi, S. Park, "A Study on the Development of an Electronic Textbook Standard in 2005," Korean Education & Research Information Service (KERIS), Research Report CR 2005-22, Republic of Korea, 2005.
8. M. Kim, K. H. Yoo, C. Park, J. S. Yoo, "Development of a Digital Textbook Standard Format Based on XML," LNCS, Vol. 6059, pp. 363–377, 2010.
9. M. Kim, K. H. Yoo, C. Park, J. S. Yoo, B. Byun, W. Cho, J. Ryu, N. Kim, "An XML-Based Digital Textbook and its Educational Effectiveness," LNCS, Vol. 6059, pp. 509–523, 2010.
10. J.-H. Hong, "Design and Implementation of 3D Digital Textbook," Master thesis, Graduate School of Education, Chungbuk National University, South Korea, 2012.
11. J.-H. Hong, M. Kim, K.-H. Yoo, "Development of a 3D Digital Textbook Using X3D," Ubiquitous Information Technologies and Applications, Vol. 214, pp. 341–351, 2013.
12. J.-U. Jeong, M. Kim, K.-H. Yoo, "Design and Implementation of 3D Annotation Techniques for 3D Digital Textbooks,", Master thesis, Graduate School of Computer Science and Education, Catholic University of Daegu, South Korea, August, 2013.
13. J.-U. Jeong, M. Kim, J.-S. Jeong, D.-H. Kim, J.-O. Kim, S.-O. Kwon, K.-H. Yoo, "Design and Implementation of 3D Annotation Features on 3D Digital Textbooks," In Proc. of Korea Computer Graphics Society, July 19–21 2013, pp. 197–198, 2013.
14. J.-E. An, Y.-B. Kim, "A Web-Based Courseware for 3D Bicycle Museum on the basis of Constructivism Learning theory," In Proc. of Korea Contents Association Spring Conference 2007, Vol. 5, No. 1, pp. 19–23, 2007.
15. H.-R., Yang, K.-K. Kang, K.-P. Han, D. Kim, "Design and Implementation of Digital Science Textbook with Cutting Effects," Journal of Korea Contents Association, Vol. 9, No. 1, pp. 465–474, 2009.
16. J.-F. Boujut, L. Gilco, "Design of a 3D annotation tool for supporting evaluation activities in engineering design," Cooperative Systems Design, Vol. 6, pp. 1–8, 2006.
17. D. Bourguigno, M.-P. Cani, G. Drettakis, "Drawing for Illustration and Annotation in 3D," Eurographics, Vol. 20, No. 3, 2001.
18. E. Y.-L. Do, M. D. Gross, "As if You Were Here – Intelligent Annotation in Space: 3D Sketching as an Interface to Knowledge-Based Design Systems," AAAI Fall Symposium, FS-04-06, pp. 55–57, 2004.
19. Sookmyung Women's University, "A Study on Sketch-based Interfaces for 3D Object Manipulation and Annotation Techniques," Research Report of ITFIND, 2006.
20. E.-J. Sin, Y.-C. Choy, S.-B. Lim, "A Study on Display Technique of Sketch Annotation in 3D Virtual Space for Collaboration," Journal of Korea Multimedia Society, Vol. 12, No. 10, pp. 1466–1477, 2009.

21. Web3D Consortium, http://www.web3d.org/, http://www.web3d.org/realtime-3d/specification/all (accessed on December 2013).
22. Web3D Consortium, "What is X3D?," http://www.web3d.org/realtime-3d/x3d/what-x3d (accessed on December 2013).

Efficient Mining of Time Interval-Based Association Rules

Ki Yong Lee and Young-Kyoon Suh

Abstract Given market or log data, it is very useful to find two sets of items or events that occur frequently with a regular time interval. We call a time-dependent relationship between two itemsets a *time interval-based association rule*. Finding time interval-based association rules, however, has not been much investigated yet until now. In this paper, we propose an efficient method for finding time interval-based association rules. The proposed method transforms the original input data into a more efficient form and then utilizes the transformed data in the subsequent steps. As a result, the input/output (I/O) cost of reading the data from disk is significantly reduced. Our experiments demonstrate the efficiency of the proposed method compared with those of the existing methods.

Keywords Association rule mining · Time-interval association rule

1 Introduction

In recent years, we are witnessing a huge volume of data pouring out from marketing, financial services, and other various fields. Given such market or log data, it is often very useful to identify two sets of items or events that occur frequently with a regular time interval. In this paper, we call such a time-dependent relationship between two itemsets a *time interval-based association rule*. The traditional association rule $X \rightarrow Y$ indicates that if a customer buys a set of items X, the customer is likely to also buy a set of items Y. By comparison, a time interval-based association rule $X \rightarrow_t Y$ means that if a customer buys X, the customer is likely

K. Y. Lee
Division of Computer Science, Sookmyung Women's University, Seoul 04310, Republic of Korea
e-mail: kiyonglee@sookmyung.ac.kr

Y.-K. Suh (✉)
School of Computer Science and Engineering, Kyungpook National University, Daegu 41566, Republic of Korea
e-mail: yksuh@knu.ac.kr

W. Lee and C. K. Leung (eds.), *Big Data Applications and Services 2017*,
Advances in Intelligent Systems and Computing 770,
https://doi.org/10.1007/978-981-13-0695-2_13

to also buy Y *after* a time interval t. For example, {milk, bread} $\rightarrow_2$ {butter} implies that if a customer buys milk and bread together, the customer is likely to also buy butter *2 days later*, if the unit of time interval is the "day."

Finding time interval-based association rules, however, has not been so far much investigated yet. To find time interval-based association rules, we may use a traditional association rule mining algorithm such as the Apriori [1] or a traditional sequential pattern mining algorithm such as the Generalized Sequential Pattern (GSP) [2]. But if those algorithms are applied directly, the number of candidate rules that need to be explored increases explosively due to an explosive combinatorial space of all possible time intervals between itemsets. Hence, the input/output (I/O) cost of counting the frequency of each candidate rule increases drastically, resulting in a much longer execution time.

To address this concern, we propose an efficient method for mining time interval-based association rules. The proposed method first transforms the original input data into a more efficient form. The method then utilizes this transformed data in order to reduce both the number of candidate rules and the I/O cost of counting the frequency of each candidate rule. As a result, the I/O cost of reading the data from disk is significantly reduced. Our experiments demonstrate the efficiency of the proposed method compared with that of the existing method.

2 Proposed Method

Let D be a set of transactions, in which each transaction consists of a timestamp and a set of items purchased together. Let I be the set of all items in D. Our problem is to find time-based association rules $X \rightarrow_t Y$, where $X, Y \subseteq I$ and $\text{supp}(X \rightarrow_t Y) \geq s$. Here, $\text{supp}(X \rightarrow_t Y)$ is the occurrence frequency of Y appearing t time units after X appears, and s is a minimum support specified by the user.

To find time-based association rules, we first transform D into another form D'. Let $ts(i)$ be the set of all the timestamps of transactions in which an item $i \in I$ appears. In D', we list $ts(i)$ for each item $i \in I$. Figure 1 exemplifies D and D'.

After obtaining D', we construct L_2 directly from D', where L_2 is the set of true time interval-based association rules $X \rightarrow_t Y$, where $|X|+|Y|=2$. Let i_1 and i_2 be two different items in I. We construct from $ts(i_1)$ and $ts(i_2)$ stored in D' a bag B in which each element is $ts_2 - ts_1$, where ts_1 and ts_2 are timestamps in $ts(i_1)$ and $ts(i_2)$, respectively. Then, for each element $b \in B$, if the multiplicity of b is larger than s, we add the time interval-based association rule $\{i_1\} \rightarrow_b \{i_2\}$ into L_2. By performing this operation for all two different items i_1 and $i_2 \in I$, we can obtain L_2. Note that we don't need to construct C_1, L_1, and C_2, unlike the traditional Apriori algorithm.

Given L_2, for each $k = 3, 4, \ldots$, we find time interval-based association rules $X \rightarrow_t Y$, where $|X|+|Y|=k$. More specifically, for each $k = 3, 4, \ldots$, we first construct from L_{k-1} the candidate set C_k containing the set of candidate association rules $X \rightarrow_t Y$, where $|X|+|Y|=k$. We then find the set of true association rules L_k from

D

Transaction ID	Timestamp	Itemset
1	1	a, c
2	1	d, e
3	2	a, c
4	3	e
5	3	d
6	3	b
7	4	a
8	4	c
9	5	b, d
10	6	a
11	6	a
12	7	c, e
13	8	b, d
14	9	a, b, c

D'

Item	Timestamps
a	1, 2, 4, 6, 9
b	3, 5, 8, 9
c	1, 2, 4, 7, 9
d	1, 3, 5, 8
e	1, 3, 7

Fig. 1 An example of an input data D and its transformed data D'

C_k by counting each $X \rightarrow_t Y \in C_k$ and selecting only those with supp($X \rightarrow_t Y$) $\geq s$. We stop this operation when $L_k = \varnothing$. Finally, we return $L_2 \cup L_2 \cup \ldots \cup L_{k-1}$ as the found rules.

After obtaining L_k, we need to construct C_{k+1} from L_k if $L_k \neq \varnothing$. To construct C_{k+1} from L_k, we take the following steps: First, we pick two rules $X_1 \rightarrow_t Y_1$ and $X_2 \rightarrow_t Y_2$ from L_k that satisfy the following conditions: (1) the time interval t is the same, (2) X_1 and X_2 are the same and Y_1 and Y_2 differ by only one item, or vice versa. For such two rules, let i_1 be the item in Y_1 (or X_1) that is not in Y_2 (or X_2). Similarly, let i_2 be the item in Y_2 (or X_2) that is not in Y_1 (or X_1). Then, we add two time interval-based association rules $X_1 \rightarrow_t Y_1 \cup \{i_1\}$ and $X_1 \rightarrow_t Y_1 \cup \{i_2\}$ to C_{k+1} if X_1 and X_2 are the same, or add $X_1 \cup \{i_1\} \rightarrow_t Y_1$ and $X_1 \cup \{i_2\} \rightarrow_t Y_1$ to C_{k+1} if Y_1 and Y_2 are the same.

Finally, we describe how to construct L_k from C_k. For each rule $X \rightarrow_t Y \in C_k$, we need to count supp($X \rightarrow_t Y$). If supp($X \rightarrow_t Y$) $\geq s$, we add $X \rightarrow_t Y$ to L_k. In fact, we can easily count supp($X \rightarrow_t Y$) using D'. Let $ts_X = \cap_{i \in X} ts(i)$ and $ts_Y = \cap_{i \in Y} ts(i)$. ts_X and ts_Y represent the sets of the timestamps of transactions in which all items in X and Y appear together, respectively. Let $ts_X' = \{ts + t \mid ts \in ts_X\}$. ts_X' is obtained by adding t to each element in ts_X. Finally, if $|ts_X' \cap ts_Y| \geq s$, then we add $X \rightarrow_t Y$ to L_k, because it means that supp($X \rightarrow_t Y$) $\geq s$. In this way, we can obtain L_k from C_k very efficiently. Figure 2 shows an example of computing supp($X \rightarrow_t Y$) using D'.

Fig. 2 An example of computing supp($\{c\} \rightarrow_1 \{b, d\}$)

Computing supp($\{c\} \rightarrow_1 \{b, d\}$):

$ts(c) = \{1, 2, 4, 7, 9\}$
$ts(b) = \{3, 5, 8, 9\}$
$ts(d) = \{1, 3, 5, 8\}$

$ts_{\{c\}} = \{1, 2, 4, 7, 9\}$
$ts_{\{c\}}' = \{1+1, 2+1, 4+1, 7+1, 9+1\} = \{2, 3, 5, 8, 10\}$
$ts_{\{b, d\}} = \{3, 5, 8, 9\} \cap \{1, 3, 5, 8\} = \{3, 5, 8\}$
$|ts_{\{c\}}' \cap ts_{\{b, d\}}| = |\{3, 5, 8\}| = 3 = \text{supp}(\{c\} \rightarrow_1 \{b, d\})$

3 Experimental Results

In this section, we present the results of performance evaluation. We have compared the performance of the proposed method with that of the Apriori algorithm, which is slightly modified to be applied to our problem. To the best of our knowledge, there was no similar work done to find time interval-based association rules [3] discussed in this paper. We used a PC with Intel i7-5820 K 3.30 GHz CPU and 8 GB memory running Windows 7. We implemented the compared algorithms in Java.

For the dataset, we generated synthetic database D, where we varied two parameters: (1) N: the total number of transactions and (2) W: the maximum number of items in each transaction. The default value for each parameter is $N = 500$ and $W = 15$. As the performance metric, we measured the execution time taken in each method. Figure 3 shows the results of performance evaluation. Figure 3a, b show the performance of both methods over increasing N and W, respectively. In all cases, the performance of the proposed method is significantly better than that of the modified Apriori algorithm. The performance improvement of the proposed method is attributed to reducing the I/O cost by using a more efficient data structure.

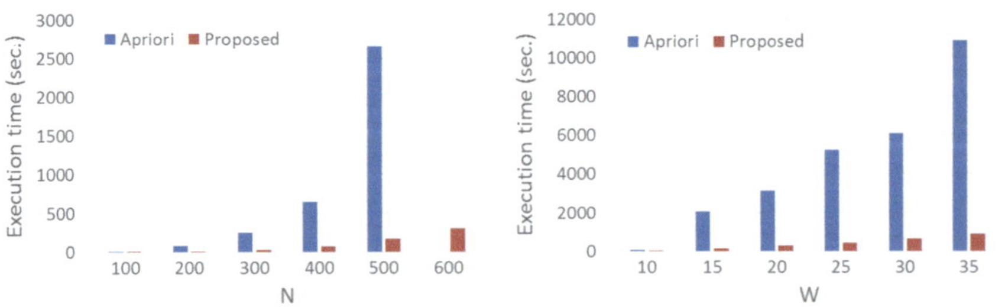

(a) Execution time comparison over increasing transaction counts

(b) Execution time comparison over increasing transaction-wise max item counts

Fig. 3 The results of performance evaluation

4 Conclusion

We introduced a new type of association rules, called time interval-based association rules, and proposed an efficient method for finding those rules. The proposed method transforms the original data into a more efficient form, resulting in substantially eliminating I/O when counting each candidate rule. In the experiments, the proposed method overwhelmed the traditional Apriori algorithm by a huge margin.

Acknowledgements This research was supported by the National Research Foundation of Korea (NRF) grant funded by the Korea government (MSIP) (No. 2015R1C1A1A02037071).

References

1. Agrawal, R., Srikant, R.: Fast algorithms for mining association rules, In: 20th International Conference on Very Large Data Bases, pp. 487–499, Santiago, Chile (1994)
2. Srikant, R., Agrawal, R.: Mining Sequential Patterns: Generalizations and Performance Improvements, In: 5th International Conference on Extending Database Technology, pp. 3–17, Springer-Verlag, London (1996)
3. Karthikeyan, T., Ravikumar, N.: A survey on association rule mining. Int'l Journal of Advanced Research in Computer and Communication Engineering, 3(1), 5223–5227, (2014)

Investigating the Role of Musical Genre in Human Perception of Music Stretching Resistance

Jun Chen and Chaokun Wang

Abstract In Big Data Era, stretching a music piece to a given length is a common demand in people's daily lives, e.g., in audio–video synchronization and animation production. However, it is not always guaranteed that the stretched music piece is acceptable for general audience since music stretching suffers from people's perceptual artifacts. Overstretching a music piece will make it uncomfortable for human psychoacoustic hearing. The research on music stretching resistance attempts to estimate the maximum stretchability of music pieces to further avoid overstretch. It has been observed that musical genres can significantly improve the accuracy of automatic estimation of music stretching resistance, but how musical genres are related to music stretching resistance has never been explained or studied in detail in the literature. In this paper, the characteristics of music stretching resistance are compared across different musical genres. It is found that music stretching resistance has strong intra-genre cohesiveness and inter-genre discrepancies in the experiments. Moreover, the ambiguity and the symmetry of music stretching resistance are observed in the experimental analysis. These findings lead to a new measurement on the similarity between different musical genres based on their music stretching resistance. In addition, the analysis of variance (ANOVA) also supports the findings in this paper by verifying the significance of musical genre in shaping music stretching resistance.

Keywords Music stretching resistance · Musical genre · Human perception

1 Introduction

Music stretching resistance (*abbr.* MSR) which describes the acceptable range of time stretching rate of music pieces for people's psychoacoustic hearing [1, 2], consists of the minimum compressing rate, denoted as α_{min} ($0.0 < \alpha_{min} < 1.0$), and

J. Chen · C. Wang (✉)
School of Software, Tsinghua University, Beijing 100084, China
e-mail: chaokun@tsinghua.edu.cn

J. Chen
e-mail: chenjun082@gmail.com

W. Lee and C. K. Leung (eds.), *Big Data Applications and Services 2017*,
Advances in Intelligent Systems and Computing 770,
https://doi.org/10.1007/978-981-13-0695-2_14

the maximum elongating rate, denoted as α_{max} $(1.0 < \alpha_{max} < 2.0)$ [3]. In Big Data Era, the research on MSR is of broad interest in fields like timescale modification of speech [1, 2], music resizing [4–6], and dynamic music rescaling [7] as well as other fields related to human perception of music and psychoacoustic hearing. The computational method [3] which estimates MSR by incorporating sound features (e.g., spectral analysis, timbre, and pitch) and musical genres has shown that musical genre is much more important in affecting MSR compared with sound features like timbre, pitch, and rhythm. However, there is still no in-depth research to investigate the relationship between MSR and musical genres to the best of our knowledge.

Generally speaking, the existence of MSR can be attributed to the process of receiving and recognizing accelerated and decelerated sounds where people would make positive or negative reaction to accept/reject the changes based on the satisfaction of their psychoacoustic hearing. Basically, MSR is related to the perception of music and the artifacts of digital signal processing. Human perception of music has much in common with human recognition of natural languages, where the structures (syntax/harmony), the vocabularies (words/chords), the tonal properties (inflection/timbre), and the temporal features (prosody/rhythm) are shared [8]. Meanwhile, the functional magnetic resonance imaging (fMRI) shows greater neuronal activity in the voice-selective regions of participants when they are listening to vocal sounds than to nonvocal sounds [9], which indicates that people may also have different degrees of sensitivity toward music with/without lyrics, or with sparse/dense lyrics. This difference may contribute to the creation of MSR since stretching operations change the density of lyrics (e.g., words per minute), too. The compressing or elongating operations [1, 2, 4–6] on a given music piece change the tempos while preserving the pitch features [1, 2, 10], which leads to a range of acceptability on user-preferred speeds [11] as well as user tolerable speed ranges. These are the evidences why MSR exists for general audience.

The dynamic attending theories [12–15] posit that the internal mechanism, like a clock, a timekeeper, or an oscillator inside human beings, resonates and synchronizes with the periodicity of stretched music pieces to enable the audience to follow changes in tempo. Without the knowledge of MSR, however, perceptual artifacts [6] are more likely to occur and degrade the auditory experience of general listeners when a music piece is stretched at a rate out of its acceptable stretching range, i.e., overly compressed or overly elongated. Most listeners are likely to identify the base tempo of music around 120 BPM (beats per minute) and the acceleration or the deceleration in speed usually induces more ambiguity in the identification of tempo [16, 17]. To some extent, the naïve approximation of acceptable stretching range, i.e., $\pm 20\%$ (namely $\alpha_{min} = 0.80, \alpha_{max} = 1.20$) [18], is baseless and fails to utilize the features of music pieces, considering that the perception of tempo appropriateness for a given music piece is determined by its contents [19]. One of the categorical features of music is genre which incorporates cultural backgrounds and emotions of artists, and characterizes the similarity between music pieces [20]. Meanwhile, musical genres are also related to music content features, e.g., timbre, pitch, and rhythm, so that the automatic genre classification algorithms [20–23] could work, which makes genre a potential factor in studying MSR.

In the previous work [3], MSR values are discretized as labels along the axis of stretching rate. The estimation of MSR is performed by classifying the label of the given music piece using its sound features (spectral analysis, timbre, pitch, and tempo) and musical genre with the machine learning techniques. It is observed that musical genre has larger contribution to improve the classification accuracy compared with sound features. But how musical genre is related to MSR has never been studied in detail. We believe that it is necessary to further explore the relationship between MSR and musical genres.

In this paper, we investigate the important role of musical genres in shaping MSR. We find that MSR tends to be constant with small fluctuation within a given musical genre. The significance of musical genres in the analysis of variance substantiates the existence of inter-genre discrepancy and intra-genre cohesiveness in MSR, i.e., MSR values are widely discrepant among different genres and are inherently cohesive within a same genre. The ambiguity of MSR is also diversified for different musical genres. Besides, the MSR values of a given musical genre are symmetric in terms of the range boundaries (i.e., mean) as well as the ambiguity of range boundaries (i.e., standard deviation). The regression lines from original tempos of music pieces to MSR values (α_{min} and α_{max}) are almost horizontal within a given musical genre in the experiments, which further indicates that MSR has little correlation with tempo unlike that with musical genre. Besides, we also analyze the MSR-based similarity between musical genres by computing the overlapping area of covering regions on the $\alpha_{min} - \alpha_{max}$ coordinate system. This new measurement of similarity offers a new perspective on musical genres based on human perception of music stretching resistance. MSR is a psychoacoustic reflection of human perception of music [11], and the study on MSR also sheds new light on content-aware music adaption [4–6] and dynamic attending theory [12–15].

2 Methods

2.1 Participants

We recruited 17 college students as participants in the experiments. These participants ranged in age from 18 to 25 years old, with 29.4% female and 70.6% male. Most of them were nonmusicians except for two participants who had received piano education for more than 3 years by the time of experiments. It has been proved that listeners (musicians or nonmusicians) can make consistent judgments on whether music pieces are played overly fast or overly slow [8, 19]. The composition of participants in our experiments is similar to the real situation in our daily lives. The participants were selected among the people who enjoyed listening to music and were willing to spend at least half an hour every day to conduct the experiments. The experiments last for about 1 month so that the experimental results would be less influenced by the short-term changes of participants' physical or mental states, such as moods (happy, sad), time (morning, evening), locations (home, workplace), and weather (rainy, sunny). Since the participants were expected to conduct the listening experiments for a long period of time, we excluded those short-term participants who could not make through the 1-month experiments to get the results from the final 17 participants. Although the final number of participants is not very large, we tried to minimize the impact of personal preferences by increasing the overlap of music pieces that different participants listened, and majority voting the results they reported. All the participants were paid a little bit for their endeavor when the experiments finished.

2.2 Experimental Settings

We used the collection of 894 songs from [3] in this study. These songs identically cover 11 musical genres as shown in Table 1. These songs were randomly crawled from a music website,[1] and the genres of these songs were annotated by referring to the metadata of these songs as well as the genre taxonomy from Wikipedia and some popular music websites.

All the songs in the collection were stretched in time domain using the synchronized-overlap-add (SOLA) method [24] to avoid pitch shift, which was implemented in the SoundTouch library.[2] Each song was stretched into different versions of discrete stretching rates between (0.00, 2.00) with a rate step 0.02. That is, each song has 49 compressed versions with stretching rates in $\{0.98, 0.96, \ldots, 0.02\}$ as well as 49 elongated versions with stretching rates in $\{1.02, 1.04, \ldots, 1.98\}$. Please note that no music piece with stretching rate beyond 2.00 will be acceptable from our experience since the elongation would have destroyed the musical structure too much and made the elongated music piece sound uncomfortable.

Although there are some other different music stretching methods [4–7], we choose SOLA [24] in this work because SOLA is a more fundamental work in the music stretching literature and it uniformly stretches music pieces. Therefore, the results reported in this paper are about uniformly music stretching.

2.3 Procedure

Each participant was delivered at least one package of 20 random songs from our collection as well as the 98 stretched versions (49 compressed + 49 elongated) for each song. The genre composition in each song package is random, and some participants were delivered more song packages since they were faster in conducting the listening experiments. The participants were asked to listen to these stretched ver-

[1] http://www.top100.cn.

[2] http://www.surina.net/soundtouch.

Table 1 The means and deviations of the MSR values of different musical genres, as well as the slopes of regression lines from the original tempo of a music piece to $\alpha_{min}/\alpha_{max}$ within each musical genre.

Genre	#.Piece	α_{min}	α_{max}	Slope α_{min}	Slope α_{max}
Pop	87	0.735 ± 0.069	1.246 ± 0.066	2.78×10^{-4}	-5.30×10^{-5}
Rock	95	0.742 ± 0.083	1.235 ± 0.064	6.85×10^{-6}	-1.45×10^{-4}
Easy Listening	83	0.594 ± 0.097	1.355 ± 0.086	-3.15×10^{-4}	6.75×10^{-5}
Folk	80	0.698 ± 0.057	1.283 ± 0.056	-4.29×10^{-5}	-1.47×10^{-4}
Latin	68	0.702 ± 0.066	1.287 ± 0.056	-1.34×10^{-4}	-1.25×10^{-5}
Country	86	0.692 ± 0.047	1.294 ± 0.050	1.73×10^{-6}	1.79×10^{-4}
Hip-hop&Rap	82	0.789 ± 0.036	1.207 ± 0.036	6.98×10^{-5}	3.12×10^{-6}
R&B	90	0.729 ± 0.045	1.259 ± 0.040	-2.77×10^{-4}	-1.71×10^{-5}
Jazz&Blues	78	0.703 ± 0.049	1.282 ± 0.059	-4.11×10^{-5}	-1.24×10^{-4}
Classical	73	0.578 ± 0.085	1.376 ± 0.079	-4.50×10^{-4}	4.89×10^{-4}
Electronic	72	0.610 ± 0.090	1.346 ± 0.076	-7.79×10^{-5}	5.31×10^{-5}

sions one after another, and judge whether each stretched version is acceptable. They could choose the order of stretching rates to listen in their preferred ways as long as they could give judgement on α_{min} and α_{max}. We offered the participants a few general judging criteria, e.g., speed acceptance, lyrics density acceptance, and overall listening acceptance. The degradation of sound quality after stretch using SOLA method also turns out to be an important factor to influence people's psychoacoustic acceptance. For example, over elongation will make the sound interruptive while overcompression will mess up the audio signals and make it sound noisy. Besides the given judging criteria, it is up to the participants to make their own judgement based on the comfort they feel in the listening. Though the total number of stretched versions of each song was not small, the participant did not have to listen to all of them, and they only needed to locate the minimum and the maximum acceptable stretching rates of each song in their package(s). To further help the participants to judge the acceptance of the stretched versions, we also developed a new music player for them, which supported "one-click" switch between two stretched versions of a same song at the same position, e.g., 30% of the song [25]. With this music player, it would be easier for the participants to judge whether the listening version is overstretched at the currently playing segment compared with the original version or the other acceptable versions.

When the participants completed the tasks in the delivered package(s), (s)he inputs the results, i.e., the α_{min} and the α_{max} of each song in the package, into our experimental web pages. Specifically, the participants enter the values of α_{min} (e.g., 0.82) and α_{max} (e.g., 1.26) of the songs in their package(s) through text fields on the web page. The participants were allowed to correct their results before the final submission by the end of the 1-month experiment. Thus, the minimum compressing rate α_{min} and the maximum elongating rate α_{max} of each song in our collection were obtained for later analysis. We have made this dataset public available.[3]

3 Results and Discussion

We performed the analysis of variance (ANOVA) on the MSR values obtained from the listening experiments. MSR values represent α_{min} and α_{max} of each song reported by the participants. The results show that MSR values are significantly affected by musical genres (F-test, alpha levels are the 11 genres, α_{min}: $F(10, 883) = 74.683$, $P < 0.001$, α_{max}: $F(10, 883) = 56.407$, $P < 0.001$). Next, the relationship between MSR and musical genres is discussed in detail based on the experimental results.

3.1 Basic MSR Properties

Figure 1 illustrates the error bars (means and standard deviations) of MSR values for the 11 musical genres (the statistics are shown in Table 1), from which we can draw the following four basic conclusions about MSR:

- **Inter-genre Discrepancy**: Wide discrepancies in MSR values under different genres are observed. The position and the stretch of MSR of musical genres are quite different. For instance, *Easy Listening*, *Electronic*, and *Classical* music have a wider stretching range (the interval between the mean of α_{min} and that of α_{max}), while *Hip-hop&Rap* and *R&B* exhibit a narrower stretching reach. It is in line with the character effect (fast and slow) of music pieces on human preferred speeds and ranges of acceptability [11].
- **Intra-genre Cohesiveness**: MSR values are substantiated to be cohesive under a given genre according to the significance of musical genres in the analysis of variance.
- **Ambiguity**: The ambiguity of MSR is greater for *Easy Listening*, *Electronic*, and *Classical* music pieces seen from the standard deviations. The larger the standard deviation is, the larger ambiguity the MSR of a musical genre is. This probably results from the rhythmic features of these musical genres since there may hardly be any fixed rhythmic patterns for these aforementioned genres, for instance, a piece of piano music or violin music. On the contrary, songs like *R&B*, *Hip-hop&Rap* usually follow a solid tempo throughout the whole music piece.
- **Symmetry**: Under a given genre, MSR tends to be symmetric between α_{min} and α_{max}, on both the range boundaries (mean) and the ambiguity of range boundaries (standard deviation). This property comes from the symmetric criteria that listeners used to judge α_{min} and α_{max}. Supposedly, the optimal tempo of a given song should

[3] https://github.com/chenjun082/msr.

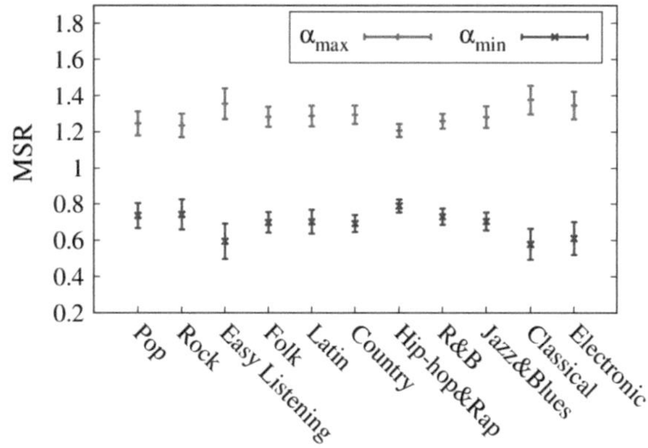

Fig. 1 The error bars of MSR distribution of 11 different musical genres

occur near its base one. As a result, if a song is over-compressed and sounds uncomfortable, it is more likely that the elongated one with the same shift of stretching rate increase will also sound uneasy, and vice versa.

3.2 *Intra-genre Linear Regression with Tempo*

The stretching operations on a given music piece will lead to an inversely proportional relationship, $t_s = \frac{t_o}{r}$, where r is the stretching rate between the range (0.00, 2.00), t_o and t_s are the tempos of the original music piece and that of the stretched version measured by beat per minute, respectively. Evidently, t_o is fixed for a given music piece. Thus, t_s meets the upper limit when r equals α_{min}, while t_s reaches the floor boundary when r equals α_{max}. So as to identify the relationship between boundary tempos and MSR, the linear regression is performed from base tempos to α_{min} and α_{max} in the music collection, respectively. This is to study whether or not songs with different tempos under a given musical genre would generally have different α_{min} and α_{max}. The regression lines are almost horizontal under all genres since their slopes are very close to zero (Table 1). For example, the steepest slope in Table 1 is -4.89×10^{-4} of α_{max} of *Classical Music*. Since the original tempo of a music piece mostly varies between [0, 200] BPM, which can only cause less than 0.1 bias in the α_{max} from other music pieces of this genre. Consequently, for a given musical genre, music pieces with different base tempos usually have similar MSR, which verifies the intra-genre cohesiveness of MSR from another point of view. Under a given genre, the limits of r are fixed, and thus the upper and the floor boundaries of t_s vary according to the value of t_o of a given music piece. The fact that the regressed slopes within each musical genre are almost zero is also a solid proof that MSR has little correlation with tempo unlike that with musical genre.

3.3 MSR-Based Musical Genre Similarity

To further investigate the relationship between MSR and musical genres, we illustrate the scatter plots of the MSR values within each musical genre in Fig. 2. Each point in these panels represents a pair (α_{min}, α_{max}), while each point may correspond to a few songs sharing the same MSR values. Obviously, the covered region of the points within each musical genre diversifies. For example, the points in Fig. 2d cover a wide range which means the ambiguity/variance of MSR in this group is high; however, the points in Fig. 2h are very close with each other spreading in a small range, and thus the ambiguity/variance of MSR in this group is relatively lower in contrast. The assembly of these points in Fig. 2a also shows the difference in the MSR value distribution of different musical genres, which inspires us that we can distinguish different musical genres based on their covering area of the points on the $\alpha_{min} - \alpha_{max}$ coordinate system.

We use a rectangle to mark the edges of the MSR region (a.k.a. MSR rectangle) on the $\alpha_{min} - \alpha_{max}$ coordinate system for a given genre. The rectangle is located by the following four coordinates:

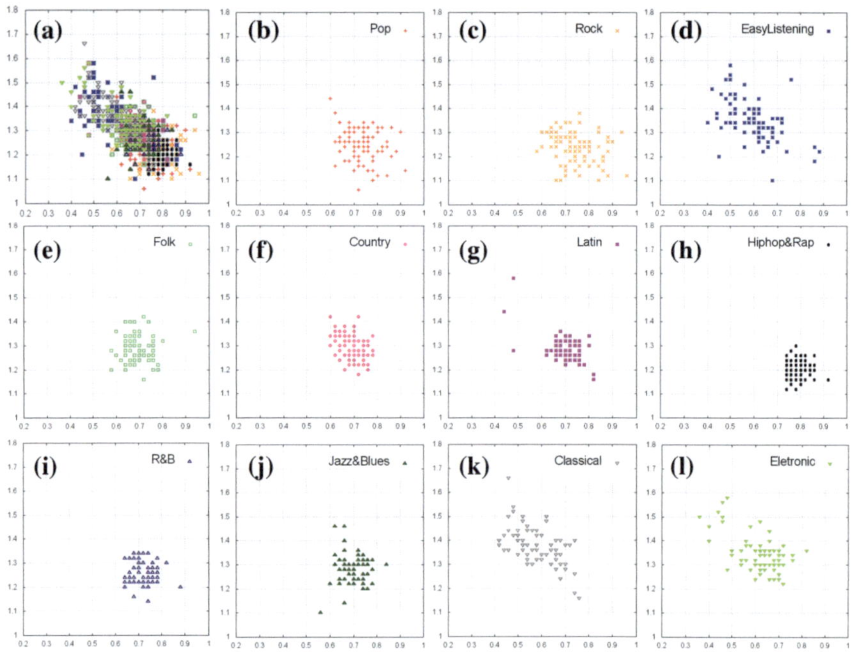

Fig. 2 The scatter plots of the MSR value distribution within each musical genre. The abscissa and the ordinate of each point in the panels represent the values of α_{min} and α_{max}, respectively. **a** is the assembly of the points from **b** to **l**

$$(Mean(\alpha_{min}) \pm StdDev(\alpha_{min}), Mean(\alpha_{max}) \pm StdDev(\alpha_{max})), \quad (1)$$

where $Mean(\alpha_{min})$ and $StdDev(\alpha_{min})$ represent the mean and the standard deviation of α_{min} in the given musical genre, respectively. So are $Mean(\alpha_{max})$ and $StdDev(\alpha_{max})$. These values are shown in Table 1. Figure 3a shows the MSR rectangle of *Pop* music. The larger the MSR rectangle is, the more ambiguity of MSR the given musical genre has. The MSR rectangle divides the $\alpha_{min} - \alpha_{max}$ coordinate system into nine parts on the first quadrant as illustrated in Fig. 3a. Part 9 can be considered as the "safe" stretching area of the given musical genre since the points in this region are within the acceptable stretching range. Parts 1, 2, 3, 4, and 7 are the "dangerous" stretching areas of the given musical genre by contrast. Furthermore, Parts 5, 6, and 8 can therefore be considered as the transition areas from the "dangerous" stretching region to the "safe" stretching region. If the music stretching tasks [1, 2, 4–6] are performed within the "safe" region, the stretched results will be the most likely to be accepted by general audience. On the contrary, if stretched in the "dangerous" region, the results will usually be unacceptable for general audience. Since different musical genres lead to different MSR rectangles, these MSR rectangles can be used to distinguish different musical genres from the perspective of MSR which has never been studied in the literature.

The area of MSR rectangle, i.e., $4 \times StdDev(\alpha_{min}) \times StdDev(\alpha_{max})$ can be used to measure the relative MSR ambiguity of musical genres. Moreover, the relationship of MSR rectangles from different musical genres falls into three categories: (1) **Inclusion**—one rectangle is totally included in another rectangle, or two rectangles are exactly the same; (2) **Exclusion**—two rectangles are totally disjoint; (3) **Intersection**—two rectangles have the common overlap as well as the disjoint part. Figure 3b shows an example of the possible relationship between musical genres based on their MSR rectangles. *Rect* 1 is intersected with *Rect* 2, while *Rect* 3 is included in *Rect* 2 and excluded from *Rect* 1. The overlaps of rectangles are filled with black, and we can compute the overlap ratio using the Jaccard similarity:

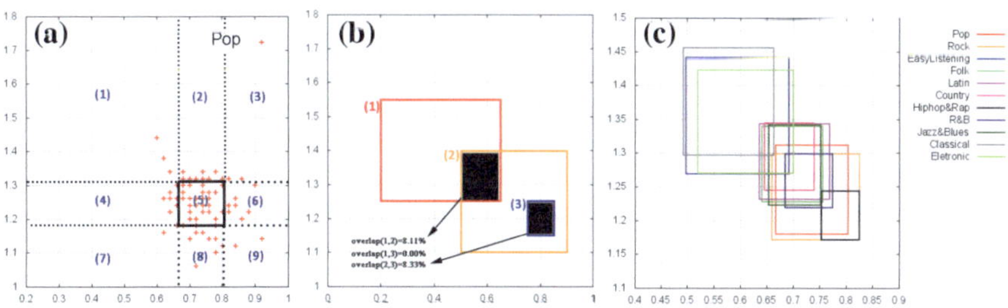

Fig. 3 The MSR-based similarity of musical genres. **a** The MSR rectangle of *Pop* music containing points on the $\alpha_{min} - \alpha_{max}$ coordinate system. **b** An example of the possible relationship between MSR rectangles. **c** The illustration of the MSR rectangles of all musical genres

Table 2 The MSR-based similarity between musical genres. The similarity matrix is symmetric, and only half the matrix is shown. The inclusion relationship is in **bold** mode and the exclusion relationship is in *italic* mode. The others are in the intersection relationship.

Genre	Pop	Rock	Easy Listening	Folk	Latin	Country	Hip-hop&Rap	R&B	Jazz&Blues	Classical	Electronic
Pop	**1.0**	0.713	0.021	0.323	0.334	0.219	0.159	**0.395**	0.346	*0.0*	0.032
Rock		**1.0**	0.018	0.255	0.259	0.168	**0.244**	**0.339**	0.275	1.7e-4	0.025
Easy Listening			**1.0**	0.082	0.092	0.088	*0.0*	0.005	0.063	0.658	0.748
Folk				**1.0**	0.808	0.675	0.002	0.344	0.822	0.024	0.113
Latin					**1.0**	0.625	0.009	0.351	0.692	0.031	0.125
Country						**1.0**	*0.0*	0.223	0.648	0.024	0.125
Hip-hop&Rap							**1.0**	0.042	*0.0*	*0.0*	*0.0*
R&B								**1.0**	0.380	*0.0*	0.014
Jazz&Blues									**1.0**	0.010	0.092
Classical										**1.0**	0.492
Electronic											**1.0**

$$Sim(Rect_i, Rect_j) = \frac{Rect_i \wedge Rect_j}{Rect_i \vee Rect_j}. \quad (2)$$

$Sim(Rect_i, Rect_j)$ measures how much similarity the musical genres that $Rect_i$ and $Rect_j$ represent share from the perspective of MSR. The relationships of MSR rectangles of all musical genres are shown in Fig. 3c. We can see that all the inclusion, the exclusion, and the intersection relationship exist among the musical genres presented in this paper.

Next, we computed the MSR-based similarity between musical genres using Eq. 2, and the results are shown in Table 2. Since the similarity matrix is symmetric, only half the matrix is shown. Obviously, the similarity between two musical genres whose MSR rectangles satisfy the *exclusion* relationship is zero (in *italic* mode). The *inclusion* relationship (in **bold** mode) is found that *R&B* $\subseteq$ *Pop*, *R&B* $\subseteq$ *Rock*, and *Hip-hop&Rap* $\subseteq$ *Rock* where $\subseteq$ is the inclusion operator. Most of the relationship observed is *intersection*. However, the ratio of overlap in the intersection relationship varies between different pairs of musical genres. The MSR-based similarity offers a new look to explore the relationship between musical genres. We can see from Table 2 that high MSR-based similarity is observed between some musical genre pairs, e.g., *Pop-Rock*, *Folk-Latin*, and *Folk-Jazz&Blues*. This kind of similarity may be very difficult or even impossible to study using metadata, audio content, or cultural backgrounds, since MSR-based similarity is related to human psychoacoustic perception of music as well as people's degrees of self-adaption to changes of music such as tempo, event density, and lyrics density.

We believe that our work has made prospective attempts in studying the relationship between MSR and musical genres, and more issues remain to be further investigated in the future. Our findings on the MSR hold not only on digital music recordings which are stretched using complex signal processing algorithms [1, 2, 4–6]. It can also be applied in live music performance, for example, the acceleration and the deceleration of live piano or violin performances to accompany singers in a concert.

Acknowledgements This work was supported in part by the National Natural Science Foundation of China (No. 61373023) and Intelligent Manufacturing Comprehensive Standardization and New Pattern Application Project of Ministry of Industry and Information Technology (Experimental validation of key technical standards for trusted services in industrial Internet). We would like to thank all the volunteers who participated in the listening experiments for their contributions which form the basis of this paper.

References

1. Verhelst, W., Roelands, M.: An Overlap-Add Technique based on Waveform Similarity (WSOLA) for High Quality Time-Scale Modification of Speech. In: IEEE International Conference on Acoustics, Speech and Signal Processing. pp. 554–557 (1993)
2. Verhelst, W.: Overlap-Add Methods for Time-Scaling of Speech. Speech Communication 30(4), 207–221 (2000)

3. Chen, J., Wang, C.: Automatic Music Stretching Resistance Classification Using Audio Features and Genres. IEEE Signal Processing Letters 20(12), 1249–1252 (2013)
4. Liu, Z., Wang, C., Wang, J., Wang, H., Bai, Y.: Adaptive Music Resizing with Stretching, Cropping and Insertion. Multimedia System 19(4), 359–380 (2013)
5. Liu, Z., Wang, C., Bai, Y., Wang, H., Wang, J.: MUSIZ: A Generic Framework for Music Resizing with Stretching and Cropping. In: Proceedings of the ACM Multimedia. pp. 523–532 (2011)
6. Liu, Z., Wang, C., Guo, L., Bai, Y., Wang, J.: Lydar: A Lyrics Density based Approach to Non-Homogeneous Music Resizing. In: IEEE International Conference on Multimedia and Expo. pp. 310–315 (2010)
7. Charles, J., Wenner, B.: Scalable Music: Automatic Music Retargeting and Synthesis. Eurographics 32(2), 345–354 (2013)
8. Brennan, D., Stevens, C.: The Effect of Pitch, Tempo and Proportional Pitch and Tempo Manipulation on Memory of Familiar Musical Excerpts. In: International Conference on Music Perception and Cognition. pp. 1771–1778 (2006)
9. Berlin, P., Zatorre, R., Lafaile, P., Ahad, P., Pike, B.: Voice-Selective Areas in Human Auditory Cortex. Nature 403, 309–312 (2000)
10. Madison, G., Paulin, J.: Ratings of Speed in Real Music as a Function of both Original and Manipulated Beat Tempo. Journal of the Acoustical Society America 128(5), 3032–3040 (2010)
11. Bisesi, E., Vicario, G.B.: Psychoacoustic Aspects of the Speed of Melody Performance. In: International Conference of Students of Systematic Musicology. pp. 7–11 (2008)
12. Large, E., Palmer, C.: Perceiving Temporal Regularity in Music. Mathematical Behavior 26(1), 1–37 (2002)
13. Jones, M., Boltz, M.: Dynamic Attending and Responses to Time. Psychological Review 96(3), 459–491 (1989)
14. Drake, C., Jones, M., Baruch, C.: The Development of Rhythmic Attending in Auditory Sequences: Attunement, Referent Period, Focal Attending. Cognition 77, 251–288 (2000)
15. Large, E., Jones, M.: The dynamics of attending: how people track time-varying events. Psychological Review 106(1), 119–159 (1999)
16. Moelants, D., Mckinney, M.F.: Tempo Perception and Musical Content: What Makes a Piece Fast, Slow or Temporally Ambiguous? In: International Conference on Music Perception and Cognition. pp. 558–562 (2004)
17. Mckinney, M.F., Moelants, D.: Deviations from the Resonance Theory of Tempo Induction. In: International Conference on Interdisciplinary Musicology. pp. 124–125 (2004)
18. Lee, E., Nakra, T.M., Borchers, J.: You're the Conductor: A Realistic Interactive Conducting System for Children. In: International Conference on New Interfaces for Musical Expression. pp. 68–73 (2004)
19. Quinn, S., Watt, R.: The Perception of Tempo in Music. Perception 35, 267–280 (2006)
20. Scaringella, N., Zoia, G., Mlynek, D.: Automatic Genre Classification of Music Content: A Survey. IEEE Signal Processing Magazine 23(2), 133–141 (2006)
21. Li, T., Ogihara, M., Li, Q.: A Comparative Study on Content-Based Music Genre Classification. In: Proceedings of the ACM SIGIR. pp. 282–289 (2003)
22. Tzanetakis, G., Essel, G., Cook, P.: Automatic Musical Genre Classification of Audio Signals. IEEE Transaction on Speech and Audio Processing 10(5), 293–302 (2002)
23. Bagci, U., Erzin, E.: Automatic Classification of Musical Genres Using Inter-Genre Similarity. IEEE Signal Processing Letters 14(8), 521–524 (2007)
24. Roucos, S., Wilgus, A.: High Quality Time-Scale Modification for Speech. In: IEEE International Conference on Acoustics, Speech and Signal processing. pp. 493–496 (1985)
25. Chen, J., Wang, C.: RESIC: A Tool for Music Stretching Resistance Estimation. In: Proceedings of the International Conference on Multimedia Modeling. Lecture Note in Computer Science, vol. 4352, pp. 386–389. Springer, Heidelberg (2014)

Keyword-Based Metadata Modeling for Experimental Omics Data Dissemination

Kyoung Soon Hwang and Keon Myung Lee

Abstract The deployment of various high-throughput instruments in molecular biology has produced a large of volume of heterogeneous experimental omics data. It is of paramount importance to collect and organize different types of those data in a unified manner so as to disseminate those valuable data to the researchers of interest. One of the essential components in such data organization and dissemination is to maintain appropriate metadata which help understand collected data. The metadata should be defined sufficient enough to provide necessary information for the researchers who want to use them. This paper presents a metadata modeling method tailored to experimental omics data which are generated for specific biological specimens. The method first identifies the candidate keywords by analyzing the literature and then they are compared with items used in other biological metadata. The candidate keywords are examined and organized into categories so that the organized keywords provide a unified view of metadata for heterogeneous collection of experimental omics data. The metadata modeling results for experimental omics data of human tissues are expressed in database schema. The proposed method has been successfully applied to metadata modeling for experimental omics data in the Biobank of Korea Centers of Disease Control and Prevention.

Keywords Omics data · Metadata · Omics database · Metadata modeling

K. S. Hwang · K. M. Lee (✉)
Department of Computer Science, Chungbuk National University, Chungdae-ro 1, Seowon-Gu, Cheongju, Chungbuk 28644, Korea
e-mail: kmlee@cbnu.ac.kr

K. S. Hwang
e-mail: hks@cbnu.ac.kr

W. Lee and C. K. Leung (eds.), *Big Data Applications and Services 2017*, Advances in Intelligent Systems and Computing 770,
https://doi.org/10.1007/978-981-13-0695-2_15

1 Introduction

With the advances of next-generation sequence technology, various omics data have been generated such as genomes, transcriptome, epigenomes, and so on. Human omics data are important assets in molecular biological studies for precision medicine. Various sequencing platforms have generated different types of omics data in different formats with different protocols. The diversity of heterogeneous types of data makes it difficult to share and integrate those omics data. Omics data are generated by sequencing instruments for biological specimens treated according to some protocol. In addition, various data processings can be applied to omics data for some purposes like noise reduction and removal, imputation, normalization, signal enhancement, quality control, and so on. The collection and management of omics data are not sufficient for their integration, sharing, and dissemination. It is necessary to collect and maintain information that explains omics data about their preparation, generation, post-processing, format, and related data and literature. Such data are called metadata which means data about data.

It is important to identify the items which are necessary to describe metadata, and to organize them into structures so that they can be effectively stored and managed. These tasks of metadata item identification and organization are called metadata modeling. Metadata modeling is a sort of engineering work whose performance is strongly affected by the skills of the modeling personnel. Therefore, there needs to be a good methodology to be followed in the course of modeling work.

This paper is concerned with a metadata modeling method for experimental omics data. We pay attention to metadata items which should be identified and collected at the moments omics data are generated, collected, and processed. We propose a keyword-based metadata modeling method for experimental omics data.

The rest of the paper is organized as follows: Sect. 2 briefly reviews related work on omics data and metadata. Section 3 presents the proposed keyword-based metadata modeling method and shows how it is applied to experimental omics data modeling. Finally, Sect. 4 draws conclusions.

2 Related Works

Several guidelines for omics data and their metadata have been proposed in the omics studies. For example, the 'Minimum information standard' of European Molecular Biology Laboratory-European Bioinformatics Institute (EMBL-EBI) [1] specifies a metadata guideline and format. As the guidelines for metadata, there are Minimum Information about a Genome Sequence/Minimum Information about a Metagenome Sequence (MIGS/MIMS) [2] for genome data, Minimal Information about a Proteomics Experiment (MIAPE) [3] for proteomic data, Minimum Information about a Microarray Experiment (MIAME) [4] for microarray data, Core Information for Metabolomics Reporting (CIMR) [5] for metabolome, and so on. As an international standardization effort, Data-Enabled Life Sciences Alliance (DELSA) was proposed, which is a checklist [6] of metadata for omics data. The checklist contains the metadata items to be managed such as the experiment institution, researchers, design, method, and data processing concerning the omics data.

There are many omics databases which archive some types of omics data with their metadata. As the representative ones, there are GenBank [7] and EMBL-EBI of genomes, Protein Information Source (PIR) [8], Protein Data Bank (PDB) [9], and Model Organism Protein Expression Database (MOPED) [10] of proteomes, Gene Expression Omnibus (GEO) [11], and ArrayExpress of transcriptomes. Sequence Read Archive (SRA) [12] is a database that stores the original sequence data produced from the next-generation sequencing technology, and contains metadata about experiments, experimental libraries, experimental protocols, gene resources, and so on. The Korea Biobank [13] provides the researchers with the information about the human body derivatives, their clinical and epidemiological data, and omics data.

There are various platforms from several vendors to produce omics data, each of which has their own data formats and standardization strategies. Although some metadata can be provided by the sequencing platforms, some metadata should be collected and archived by human personnel involved in biological material preparation, omics data recording and registration, and omics data preprocessing.

There have been some works to metadata modeling in omics databases such as The Cancer Genome Atlas (TCGA) [14], Database of Genotypes and Phenotypes (dbGaP) [15], and Database of Genomic Variants archive (DGVa) [16] and GEO. Each database has its own purpose of their services and its own scope of data and services. We pay attention to the integration of heterogeneous experimental omics data and dissemination of the omics data along with plentiful metadata to describe the omics data.

Generating omics data is costly and complicated because sequencing needs to be done through several complicated processings and expensive instruments are used. In order for third-party researchers to use data produced by others, they need to know all details of the data generation process. This paper is concerned with metadata modeling which allows third-party researchers to use omics data produced by others as like their own. We have analyzed the existing metadata guidelines and several omics databases which provide some metadata. We found out that the metadata are not sufficient to provide comprehensive information for omics data that are generated for human body derivatives from which various omics data are produced. Hence, we propose a metadata modeling method for experimental omics data. The proposed method can be applied to construct metadata for other experimental data.

3 The Proposed Keyword-Based Metadata Modeling Method

Experimental omics data are generated from sequencing instruments and stored in the instrument-specific formats. Sometimes omics data are transformed into a specific standard format. Metadata are information necessary for third-party researchers to be able to use experimental omics data as if they are generated and processed by themselves. Those metadata are generated not by sequencing instruments but by persons who have been involved in the biological material preparation, sequencing, post-processing, and data collection and their archiving. It is very important to determine which items should be collected as metadata in experimental data.

In existing metadata modeling, selection, and organization of metadata items are made by the persons who are in charge of metadata modeling. The quality and completeness of metadata modeling depend on the capability and skills of the modeling persons. We propose a new metadata modeling method that can be applied to experimental data. The method consists of the following steps:

First, we enumerate all data tables for which metadata should be modeled. For the data tables, the items corresponding to attributes are categorized into data items and metadata items, if some metadata are stored in them. In the databases with no explicit metadata models, some metadata can be stored and managed in the data tables.

Second, we create the preliminary groups of metadata items with the help of the domain experts. In this step, the domain experts may provide additional metadata items which they think to be maintained for improved services. The preliminary groups are later refined with newly identified metadata items.

Third, we identify metadata items from other related databases and guidelines, and from literature. It is crucial to find out unidentified items which give meaningful information to the researchers who use the data. The proposed method pays attention to keywords which can be metadata items. The metadata items are extracted from similar or related databases by analyzing the database schema and some related metadata guidelines, if exist. The research papers and reports can be valuable sources from which metadata items are identified. The experimental data like omics data are used by the researchers and the research results are described with the sections like 'Materials and Method' in which they present what and how they have treated and processed. To identify such candidate items for metadata, we present a keyword-based method.

Fourth, we organize all identified metadata items into hierarchical structure from which the metadata schema is defined. Metadata are also a kind of data which needs to be managed in database tables. The hierarchical structure of metadata items starts from the preliminary groups obtained in the second step.

Fifth, we define the metadata schema from the hierarchical structure and its items obtained in the fourth step. For each metadata table, the items are specified by their identifier, their data type, the enforced constraints, and the annotation to explain the meaning and role of the item.

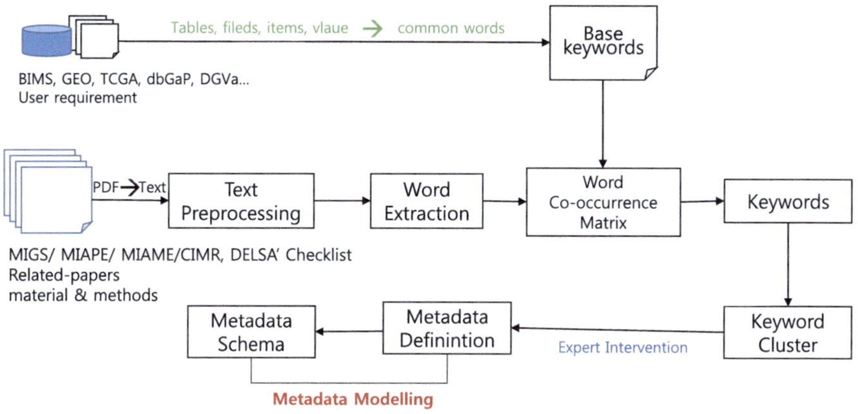

Fig. 1 The process of keyword-based metadata modeling

3.1 Keyword-Based Metadata Item Extraction

In metadata modeling, the proposed method analyzes all available research publications like papers, reports, and experiment and data processing notes in order to identify metadata items. Figure 1 shows the outline of the keyword-based metadata modeling process.

To begin with, metadata items are extracted from related databases in which data items describing the experimental data themselves are ignored, and only items describing the explanatory and management information about the data are included. The extracted items are called base keywords. To discover unidentified metadata items, the above-mentioned publications are analyzed as follows: The analysis is conducted on the section "Materials and Methods" and corresponding sections. First, with a text processing tool, all tokens are extracted and the frequent tokens are left out because they are not usually important keywords. Among the remaining tokens, frequent tokens and frequent adjacent tokens over the examined publications are extracted. The frequent tokens are examined to see if they deserve to be a metadata item with the help of domain experts. The identified keywords are first categorized according to the preliminary groups and the preliminary groups are re-organized into hierarchical structure if needed.

3.2 Keyword-Based Metadata Modeling for Experimental Omics Data

Now let us show how the proposed keyword-based metadata modeling method can be applied to experimental omics data. Here the experimental omics data assumes

Table 1 Requirements for the experimental omics metadata

		Mandatory information	Optional information
Management information	Project	ProjectID, ProjectDescription, PIName, CohortID, SubmitDate, PlatformID, …	Does not apply
	Outcomes	OutcomesID, ExperimentID, DataType PublicationTitle, DataFileLocation, …	
	Dissemination history	SampleID, ExperimentID, ApplicantID, SubmissionDate, …	
Experimental Data Information	Sample	ProjectID, CohortID, SampleNum, …	
	Experimental information	ExperimentID, ProjectID, PlatformID, ExperimentDescript, omicsDataType, …	Experimental methods conditions, S/W, …
	Data/sample QC	ExperimentID, SampleNum, scanning	Does not apply
Experimental Dataset Information	Outcomes files	OutcomesID, FileType, LocationInfor	
	Sample	CohortID, LocationInfor	
	Raw data files	ExperimentID, omicsDataType, DataLevel, PlatformID FileType, LocationInfor	

to contain genome data, transcriptome data, metabolome data, and some other Next-Generation Sequencing (NGS) data.

The first step is to enumerate all the data tables and their items, and then to distinguish metadata items from data items. In addition, we clarify the requirements on metadata description from the domain experts about the objective of metadata in the domain. Table 1 shows an example of extracted requirements from the interview with the domain experts. It describes the categories and their items to be required to be managed in the perspective of domain experts.

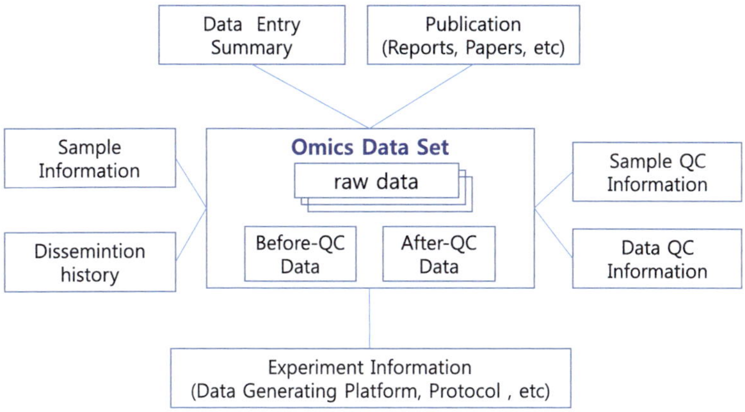

Fig. 2 The identified preliminary groups of metadata obtained from the original data tables

The second step is to group the metadata items with respect to the experimental omics data. In this step, the preliminary groups are identified along with their items which are available at the moment. Figure 2 shows the preliminary groups of metadata identified. Here the omics data are categorized into raw data, before-quality control (QC) data, and after-QC data. The identified preliminary groups are data entry summary, publications, sample information, dissemination history, sample QC information, data QC information, and experiment information.

The third step is to identify new metadata items from related or similar databases and publications. Table 2 shows some metadata items from related databases in which corresponding items are included. From this comparison, we can evaluate the importance of identified items. From the publications, the metadata items are extracted by analyzing the Materials and Method section using the template shown in Table 3 [17–21]. In the template, the "Descriptions" field contains the statements from the publications which the domain experts examine when they judge whether the identified tokens deserve to be keywords.

Step 4 is to organize all identified metadata items into hierarchical structure as shown in Fig. 3. Here the metadata items are first grouped into management information and experimental information. Then, management information is further grouped into project information, outcomes information, and dissemination information. Experimental information is also categorized into sample information and experimental data information. This kind of hierarchical organization makes it clear how the metadata items are grouped together and how to transform them into database tables.

Step 5 is to define the metadata schema from the organized structure of metadata items. Before defining the schema, the metadata items are descriptively specified with its identifier and annotation describing its meaning and role as shown in Fig. 4. The documentation like Fig. 4 should be reviewed by the domain experts for their importance as the metadata and the accuracy of the descriptions. Once the approval

Table 2 Metadata items from related databases and guideline

Items		DELSA global	TCGA	dbGAP	GEO	DGVs
Experiment (mandatory items)	Experiment ID	o	o	o	o	o
	Experiment title	o			o	
	Experiment description		o		o	o
	Platform code	o		o	o	o
	Experiment design	o	o		o	o
	Data type					
	Appendix	o				
SNP (option items)	Growth protocol	o			o	
	Treatment protocol	o			o	
	Extraction protocol	o	o		o	
	Label protocol	o	o		o	
	…					
	Imputation S/W option	o				
	Reference panel	o				
	Data processing	o	o		o	
	Data value definition	o	o		o	

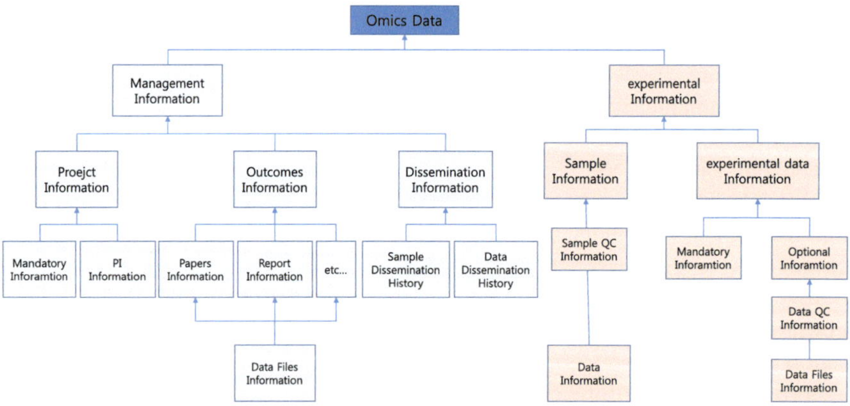

Fig. 3 Hierarchical structure of metadata items

of the metadata model is granted by the domain experts, the metadata schema is defined.

Figure 5 shows a metadata model for an experimental omics data which is expressed in the Entity Relationship (ER)-diagram. Each item is specified by its identifier, data type, and associations with other items. From the ER-diagram, the schema of metadata tables can be defined by grouping the connected items with

Table 3 An example of candidate keyword extraction from publications

Metadata		Descriptions	Paper #	Keywords
CNV (copy number variant)	Data type	Sequencing, Oligo aCGH, SNP array, BAC aCGH, PCR, qPCR	3,4,5,6	**Method type** (dbVar)
	Labeling	One of two restriction enzymes, Nsp I or Sty I, and ligated to Nsp I or Sty I adaptors./The ASD test and reference samples were labeled with Cy5 and Cy3, respectively, using Invitrogen BioPrime CGH labeling kit (Invitrogen, Carlsbad, CA). All samples have been run on Agilent 1 M CGH array according to manufacturer's protocol. DNA was labeled with Cy5-dCTP, and normal male reference DNA with Cy3-dCTP (Amersham Biosciences UK Ltd., Little Chalfont, United Kingdom), using a BioPrime **Labeling Kit** (Invitrogen)	3,6,7	**Label protocol** (GEO) **Label** (TCGA) **Nsp I, Sty I, Cy3**-dCTP, **Cy3**dCT, **Labeling Kit**
	Hybridization	In the Affymetrix Hybridization Oven 640. Hybridization was performed overnight at 37 °C. After stringent washes, the slides were stained with anti-digoxigenin-rhodamine (Roche Diagnostics) for the digoxigenin-labeled probes and embedded in an antifade solution (Vectashield, Vector Laboratories, Burlingame, CA, USA) containing 4',6-diamidino-2-phenylindole (DAPI) as counterstain	3,7	**Hyb protocol** (GEO) **Hybridization Name** (TCGA)
	Scanning	Images of the arrays were obtained using an Affymetrix GeneChip **Scanner** 3000. The arrays were scanned using the Agilent microarray **scanner**. The slides were scanned with ScanArray4000 confocal laser **scanner** (Perkin Elmer, Boston, MA)	3,4,6,7	**Scan protocol** (GEO) **scan name** (TCGA)

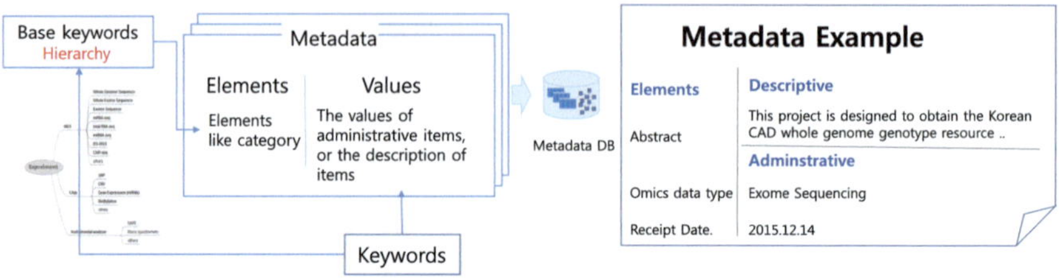

Fig. 4 The metadata structure and examples

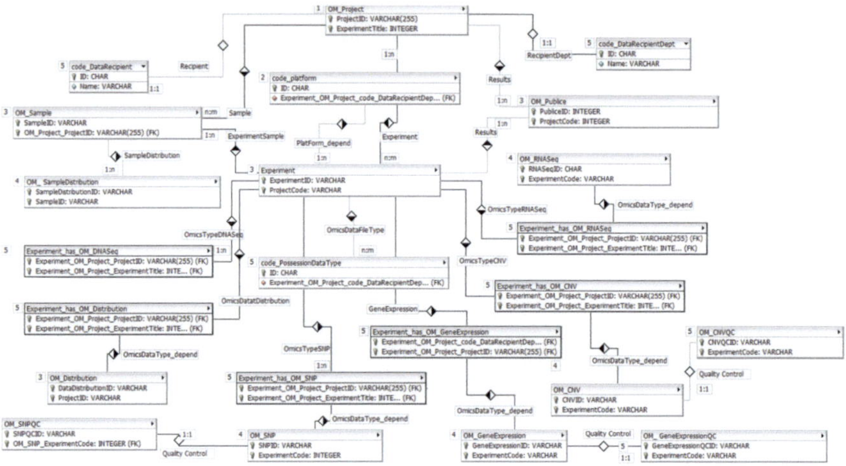

Fig. 5 A metadata model of an experimental omics data

respect to the semantic distance and storage duplication. Table 4 shows an example of a metadata table for an omics data called single nucleotide polymorphism (SNP). It lists the metadata items along with some example values for those items and the asterisk sign (*) indicates that the corresponding item is optional.

4 Conclusions

This paper proposed a keyword-based metadata modeling method with which we can systematically identify metadata items in experimental omics data and organize them. The method has been successfully applied to define metadata of experimental omics data collected and archived in the Biobank of Korea Centers of Disease Control and Prevention. From the identified metadata items, the metadata schema was defined which can be later used to build a database for metadata. If metadata items are collected and managed when omics data are generated and collected, the omics database systems can provide sufficient information for researchers to use omics data.

Table 4 An example of metadata table for SNP

Metadata		Values
Experiment (mandatory items)	Experiment ID	2009-E00000-00-001-01
	Experiment title	Affymetrix SNP array data for acute lymphoblastic leukemia samples
	Experiment description	We analyzed the Affymetrix Genome-Wide Human SNP array 6.0 and generated 906,660 individual genotypes in 3005 samples
	*Omics data type	SNP
	*Platform	Affymetrix Genome-Wide Human SNP array 6.0
	Experiment Design	Affymetrix® Genome-Wide Human SNP Nsp/Sty 6.0 User Guide.pdf
	Data type	PLINK, CHP, CEL
*SNP (option items)	Growth protocol	ES cell-derived NS cells were routinely generated by re-plating d 7 adherent neural differentiation …
	Treatment protocol	RNA was extracted with Trizol reagent, followed by clean-up and DNase I treatment with QIAGEN RNeasy mini …
	Extraction protocol	Genomic DNA was extracted from mononuclear cell preparations of leukemic blasts obtained from peripheral blood or bone marrow at diagnosis …
	Label protocol	Agilent Genomic Enzymatic labeling kit, Agilent Oligonucleotide Array-based CGH protocol Ver. 6.0, …
	Hybridization protocol	Oligo aCGH Hybridization kit, −65 °C, 24 h, 20 rpm
	Chip scan protocol	Gene Chip Scanner 3000 reference manual imager S/W
	Scan S/W	The array image was acquired using Affymetrix GeneChip® Operating Software
	…	

The identified metadata items organized into the categories can be used to collect auxiliary data for experimental omics data. The proposed keyword-based metadata modeling method is expected to be applied to other experimental databases.

Acknowledgments This work was supported by the Research Program funded by the Korea Centers for Disease Control and Prevention (Grant No.: 11-1352173-000267-01) and by Next-Generation Information Computing Development Program through the National Research Foundation (NRF) of Korea (Grant No.: NRF-2017M3C4A7069432), by the MSIT (Ministry of Science and ICT), Korea.

References

1. The European Bioinformatics Institute, https://www.ebi.ac.uk/
2. MIGen, http://www.mibbi.org/inde x.php/ projects/MIGen/
3. Minimum Information about a Phylogenetic Analysis, http://www.mibbi.org/index.php/projects/MIAPA
4. https://www.ncbi.nlm.nih.gov/geo/info/MIAME.html
5. http://mibbi.sourceforge.net/projects/CIMR.shtml
6. Snyder M, Mias G, Stanberry L, Kolker E: Metadata Checklist for the Integrated Personal Omics Study: Proteomics and Metabolomics Experiments: Vol 4, 202–206, Big Data, (2013)
7. National Center for Biotechnology Information, http://www.ncbi.nlm.nih.gov
8. The Protein Information Resource, http://pir.georgetown.edu/
9. The Protein Data Bank archive, https://www.wwpdb.org/
10. Eugene Kolker, Roger Higdon, Winston Haynes, Dean Welch, William Broomall, Doron Lancet, Larissa Stanberry, and Natali Kolke: MOPED: Model Organism Protein Expression Database: Vol. 40, Nucleic Acids Research, (2012)
11. Gene Expression Omnibus, https://www.ncbi.nlm.nih.gov/geo/
12. Sequence Read Archive Information, https://www.ncbi.nlm.nih.gov/sra
13. Biobank Information Management System, http://is.cdc.go.kr/
14. The Cancer Genome Atlas (TCGA), https://cancergenome.nih.gov/
15. The database of Genotypes and Phenotypes, https://www.ncbi.nlm.nih.gov/gap
16. Database of Genomic Variants archive, https://www.ebi.ac.uk/dgva
17. T. J. Pugh, A. D. Delaney, N. Farnoud1, S. Flibotte, M. Griffith, H. I. Li1, H. Qian, P. Farinha, R. D. Gascoyne and M. A. Marra: Impact of whole genome amplification on analysis of copy number variants: Vol. 36, No. 13, Nucleic Acids Research (2008)
18. Ingeborg B, Anette SB Wolff, Kristian L, Per M Knappskog and Eystein S Husebye : Genome-wide copy number variation (CNV) in patients with autoimmune Addison's disease: BioMed Central, (2011)
19. Shanker Swaminathan, Matthew J. Huentelman, Jason J. Corneveaux, Amanda J. Myers, Kelley M. Faber, Tatiana Foroud1, Richard Mayeux, Li Shen, Sungeun Kim, Mari Turk, John Hardy, Eric M. Reiman, Andrew J. Saykin: Analysis of copy number variation in Alzheimer's disease: the NIALOAD/ NCRAD Family Study: Vol. 9(7):801-14, pubMed, (2012)
20. Aparna Prasad, Daniele Merico, Bhooma Thiruvahindrapuram, John Wei, Anath C. Lionel, Daisuke Sato, Jessica Rickaby, Chao Lu, Peter Szatmari, Wendy Roberts, Bridget A. Fernandez, Christian R. Marshall, Eli Hatchwell, Peggy S. Eis, and Stephen W. Scherer: A Discovery Resource of Rare Copy Number Variations in Individuals with Autism Spectrum Disorder: Vol. 2 (12), G3.Genes|Genomes|Genetics, (2012)
21. X Zheng, FY Demirci, MM Barmada, GA Richardson, OL Lopez, RA Sweet, MI Kamboh and E Feingold: Genome-wide copy-number variation study of psychosis in Alzheimer's disease: Transl Psychiatry, (2015)

Nonlinear Time-series Mining of Social Influence

Thinh Minh Do, Yasuko Matsubara and Yasushi Sakurai

Abstract Given a large collection of time-evolving online user activities, such as Google Search queries for multiple keywords of various categories (celebrities, events, diseases, etc.), which consist of d keywords/activities, for l countries/locations of duration n, how can we find patterns and rules? For example, assume that we have the online search volume for "Harry Potter", "Barack Obama", and "Amazon", for 232 countries/territories, from 2004 to 2015, which include external shocks, sudden change of search volume, and more. How do we go about capturing nonlinear evolutions of local activities and forecasting future patterns? In this paper, we present Δ-SPOT, a unifying analytical nonlinear model for analyzing large-scale web search data, which is sensemaking, automatic, scalable, and free of parameters. Δ-SPOT can also forecast long-range future dynamics of the keywords/queries. We use the Google Search, Twitter, and MemeTracker dataset for extensive experiments, which show that our method outperforms other effective methods of nonlinear mining in terms of accuracy and in both fitting and forecasting.

Keywords Time-series analysis · Social influence analysis · Parameter-free

1 Introduction

Online news, blogs, SNS, and many other web search services have been speedily developing and playing a very important part in information searching. One of the goals is to find a good presentation of the data, from which we can excavate important

T. M. Do · Y. Matsubara · Y. Sakurai (✉)
Kumamoto University, 2-39-1 Kurokami, Chuo Ward,
Kumamoto Prefecture 860-8555, Japan
e-mail: yasushi@cs.kumamoto-u.ac.jp

T. M. Do
e-mail: do@dm.cs.kumamoto-u.ac.jp

Y. Matsubara
e-mail: yasuko@cs.kumamoto-u.ac.jp

W. Lee and C. K. Leung (eds.), *Big Data Applications and Services 2017*,
Advances in Intelligent Systems and Computing 770,
https://doi.org/10.1007/978-981-13-0695-2_16

patterns and properties. In other words, our goal is to detect patterns, rules, and outliers in a huge set of web search data, consisting of tuples of the form: *(query, location, and time)*. For example, assume that we have the online search activities for "Harry Potter", in 232 countries/territories from 2004 to 2015. So, how can we find meaningful information, such as the external shocks/events that happened during the period of 11 years? In the case that such events happened, do they have any relationship between each other (frequent/cyclic events or not?). Also, can we detect global/local-level patterns? Are there countries/locations that react differently from the global trend? Can we forecast the dynamics of future events?

In this paper, we propose Δ-SPOT, a unifying analytical nonlinear model which is sensemaking, scalable, and parameter-free, and provides a good summary of large collections of local online activities. Intuitively, we wish to solve the following problem:

Informal problem. Given a large collection of online activities, which consists of d keywords in l locations of duration n with complex dynamics and external shocks, we want to

- find global and local patterns,
- detect external shocks (important events in reality), and
- forecast future activities.

Especially, we want to capture all these features automatically and effectively.

Preview of our results. Figure 1a shows the search volume for the keyword "Harry Potter" from 2004 to 2015 (11 years) as gray circles, and our fitted model, as solid red line. Our method automatically spots seven big cyclic/noncyclic events that relate to "Harry Potter". We also discover two kinds of external shock events: cyclic and noncyclic. In more details,

- the biennial release (in July) of "Harry Potter" movies and books (shown as green circles), which corresponds to the major publication of works on "Harry Potter",
- the release of new episodes of "Harry Potter" movie series, held in November (shown as purple circles), and
- noncyclic shock in May (shown in red circle).

Figure 1b shows the worldwide reaction to the release of the last episode of "Harry Potter". It is clearly shown that most of the active (high reaction-level) countries have huge number of fans of "Harry Potter" (For example, the U.S., European countries, English native speaking countries, etc.)

Contributions. In this paper, we propose Δ-SPOT, a unifying analytical nonlinear model for large-scale online user activities. Briefly, our proposed model, Δ-SPOT, has the following desirable properties:

1. **Sensemaking**: Our method can detect external shocks which are related to real-time events, such as the annual sporting events, or the release of new products.
2. **Automatic**: Thanks to our modeling framework, Δ-SPOT is fully automatic, requiring no manual tuning, where the goal is to minimize the resulting modeling cost.

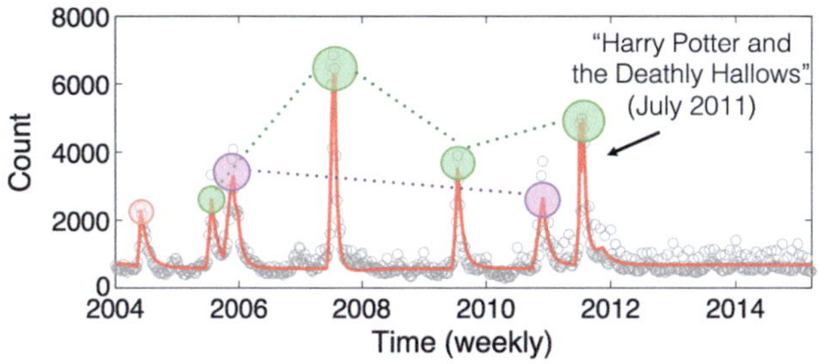

(a) Fitting result of Δ-SPOT

(b) World-wide reaction

Fig. 1 Modeling power of Δ-SPOT: **a** It automatically detects external events of keyword "Harry Potter" and **b** the worldwide reaction to the release of the last episode of "Harry Potter" movie series

3. **Scalable**: Our method scales linearly to the input size.
4. **Parameter-free**: Δ-SPOT requires no parameters or specialized tuning.

Outline. The rest of the paper is organized as follows. Next, we describe related work, followed by our proposed model, algorithms as well as experiments, discussion, and conclusions.

2 Related Work

We provide a survey of the related literature, which falls into the following categories: (a) Pattern discovery in time series and (b) social activity analysis.

Pattern discovery in time series. Traditional approaches typically use linear methods, such as auto-regression (AR), linear dynamical systems (LDS), TBATS [10], and their variants [5, 7, 8, 23]. TriMine [14] is a scalable method for forecasting complex time-stamped events, while [11] developed AutoPlait, which is a fully automatic mining algorithm for co-evolving sequences. Rakthanmanon et al. [20] proposed a similarity search algorithm for "trillions of time series" under the DTW distance.

Social activity analysis. Recently, analyses of epidemics and social media have attracted a lot of interest. The work described in [15] studied the rise and fall patterns in the information diffusion process through online social media. Prakash et al. [18] described the setting of two competing products/ideas spreading over a network, and provided a theoretical analysis of the propagation model for arbitrary graph topology. FUNNEL [16] is a nonlinear model for spatially co-evolving epidemic tensors, while EcoWeb [12] is the first attempt to bridge the theoretical modeling of a biological ecosystem and user activities on the Web. [13] developed CompCube, which operates on large collections of co-evolving activities and summarizes them succinctly with respect to multiple aspects (i.e., keyword, location, and time). For online activity analysis, Gruhl et al. [4] explored online "chatter" (e.g., blogging) activity and measured the actual sales ranks on Amazon.com. Ginsberg et al. [2] examined a large number of search engine queries tracking influenza epidemics. They described that the evolutions of search engine keywords are highly correlated with actual flu virus activity. Besides, the work reported in [1, 3, 19] studied keyword volume, to predict consumer behavior.

Contrast to the competitors. Table 1 illustrates the relative advantages of our method. Only our Δ-SPOT matches all requirements, while

- The SI model (SIR, SIRS, *SKIPS* [22], etc.) can compress the data into a fixed number of parameters and capture the dynamics of epidemiological data; however, it cannot describe periodic events and is incapable of forecasting.
- The traditional AR, ARIMA, and related forecasting methods including AWSOM [17], PLiF [9], and TriMine [14] are *fundamentally* unsuitable for our setting, because they are based on linear equations, while we employ *non*linear equations. Moreover, most of them require parameter tuning.
- FUNNEL [16] is a comfortable nonlinear model for time-evolving tensor mining. However, it cannot detect cyclic external shocks and was applied for epidemic sequences.

Table 1 Capabilities of approaches. Only our approach meets all specifications

	SI/++	AR/++	FUNNEL	Δ-SPOT
Nonlinear	✓		✓	✓
Outliers detection			✓	✓
Online activities				✓
Cyclic events/shocks				✓
Local analysis			✓	✓
Parameter-free			✓	✓
Forecasting		✓	✓	✓

3 Proposed Model

3.1 Intuition Behind Our Method

Assume that we receive time-stamped activities of the form (*query, location, and time-tick*). We then have a collection of sequences with d unique keywords, l locations/countries with duration n. We can treat this set of $d \times l$ sequences as a third-order tensor, i.e., $\mathcal{X} \in \mathbb{N}^{d \times l \times n}$, where the element $x_{ij}(t)$ of $\mathcal{X}$ shows the total number of entries of the ith keyword in the jth country at time-tick t. For example, ("Harry Potter", "US", "July 15–21, 2007", 100) means that the search volume for "Harry Potter" in "US" on "July 15–21 in 2007" is "100". We refer to each sequence of the ith keyword in the jth location: $\boldsymbol{x}_{ij} = \{x_{ij}(t)\}_{t=1}^{n}$, as a "local/country"-level web search sequence. Similarly, we can turn these local sequences into "global/world"-level web search sequences: $\bar{\boldsymbol{x}}_i = \{\bar{x}_i(t)\}_{t=1}^{n}$, where $\bar{x}_i(t)$ shows the total count of the ith keyword at time-tick t, i.e., $\bar{x}_i(t) = \sum_{j=1}^{l} x_{ij}(t)$ (Table 2).

Here, let us provide the reader with several important observations.

(P1) Basic trends. We assume that the popularity size of each activity evolves over time. The popularity size corresponds to the aggregated volume of each user who is interested in each topic in each country. For example, the release of a new movie of the series "Harry Potter" is a very big event, especially to all Harry Potter's fans all around the world. Many users will spend time searching for the new movie's information, including the release time and the cinema tickets price. Furthermore,

Table 2 Symbols and definitions

Symbol	Definition
d	Number of keywords/queries
l	Number of locations/countries
n	Duration of sequences
$\mathcal{X}$	third-order tensor ($\mathcal{X} \in \mathbb{N}^{d \times l \times n}$)
$\boldsymbol{x}_{ij}$	Local-level sequence of keyword i in location j i.e., $\boldsymbol{x}_{ij} = \{x_{ij}(t)\}_{t=1}^{n}$
$\bar{\boldsymbol{x}}_i$	Global-level sequence of keyword i i.e., $\bar{\boldsymbol{x}}_i = \sum_{j=1}^{l} \boldsymbol{x}_{ij}$
$S_{ij}(t)$	Count of (S)usceptibles i in location j at time t
$I_{ij}(t)$	Count of (I)nfectives i in location j at time t
$V_{ij}(t)$	Count of (V)igilants i in location j at time t
$\mathbf{B_G}$	Base global matrix ($d \times 4$) i.e., $d \times \{\mathbb{N}, \beta, \delta, \gamma\}$
$\mathbf{B_L}$	Base local matrix ($d \times l$) i.e., $\mathbf{B_L} = \{b^{(L)}{}_{ij}\}_{i,j=1}^{d,l}$
$\mathbf{R_G}$	Growth effect global matrix ($d \times 2$) i.e., $d \times \{t_\eta, \eta_0\}$
$\mathbf{R_L}$	Growth effect local matrix ($d \times l$) i.e., $\mathbf{R_L} = \{r^{(L)}{}_{ij}\}_{i,j-1}^{d,l}$
$\mathcal{S}$	External shock tensor i.e., $\mathcal{S} = \{\mathbf{s}_1, \mathbf{s}_2, \ldots, \mathbf{s}_k\}$
$\mathcal{F}$	Complete set of parameters i.e., $\mathcal{F} = \{\mathbf{B_G}, \mathbf{B_L}, \mathbf{R_G}, \mathbf{R_L}, \mathcal{S}\}$

they will share this event to their friends via the social network; and eventually, this leads to an exponential growth in popularity size.

(P2) Area specificity. For each topic, users all over the world have various types of reaction toward it. Due to the network connection condition, or some specified reasons, there may be a huge external shock in some countries at a specified time-tick, while nothing happens in other countries. Namely, we want to detect local-level anomalies from global-level pattern. For example, in Fig. 1b, users all over the world react differently to the release of the last episode of "Harry Potter" movie series. Mostly, users from countries, in which Harry Potter are popular, are highly active to search for the episode's information.

(P3) Population growth effect. We also find that there exists a sudden change of popularity base size in some sequences. We call this behavior the population growth effect. This phenomenon consists of different behaviors compared to the external shock effect. The dynamics of the search volume are affected not only by the events itself but also by the huge change in human demands for the topic. For example, the online service decade brings people so many innovation; as the result, the search volume for online service relating keywords increases sharply, not because of any special events. We cannot consider this phenomenon a big-scale external shock, because the behavior is different from normal shocks. We will discuss it further in Sect. 3.3 (Fig. 4).

(P4) External shock events. One of our main goals is to detect the external shocks that refer to real-time events. The search volume for a keyword sharply rises when some events (new release, tournaments, performance, etc.) relating to that keyword happened. For example, Fig. 1a shows seven big events corresponding to the release date of Harry Potter movies and books, from the first shock in 2004 (movie, episode 3), until the last one in 2011 (movie, episode 7—part 2). More importantly, we observe that some external shocks have the cyclic pattern. These cyclic events happen at the same time-tick of a specific window size (i.e., 1 year, 2 years, etc.), within the same duration. It is important to extract these cyclic patterns from the large set of all external events. In other words, we want to capture the periodical (annual, biennial, and quadrennial) events, which provides a highly sensitive view of big events in reality.

Summary. In this paper, we propose a new model, namely, Δ-SPOT, which tries to incorporate all the above important properties that we observed in the real dataset. Consequently, we would like to capture the following properties:

- **(P1)**: basic trends
- **(P2)**: area specificity and sensitivity
- **(P3)**: population growth effect
- **(P4)**: cyclic external events

3.2 *Δ*-SPOT—*With a Single Sequence*

We begin with the simplest case, assuming that we are given a single sequence. The model we propose has nodes (=users) of three classes:

- Susceptible: nodes in this class can get influenced by their neighboring nodes who have searched for it. In other words, citizens of this class are always ready to search for the keywords.
- Infective: nodes who already searched for the keywords, also capable of influencing other available nodes (share, or tell the story about the keywords), namely, transmitting the interest in the topic to the citizens in the susceptible class.
- Vigilant (i.e., busy/unavailable): nodes in this class do not get condition to search for the information (no network connection, no free time to care about the topic), so they are immune to the influence of the trend.

When an external event happens, or a controversial topic that is represented by a query rises, we assume that it "infects" all online users. Among them, many Susceptible users are vulnerable to the "infection"; hence, they immediately search, tweet, or share the information of the external events on social media. As the result, these users change their status to Infective. Also, they keep "infecting" their friendly users to search for the news due to the event's "infection". We refer to this external event as a shock, with a starting time and a specified duration. Here, the Vigilant users are immune to the "infection", so they do not have any actions during the external event. Eventually, the event comes to an end, which means the almost users drop their attention to it, as well as do not cause any "infection" to other users (Fig. 2).

Figure 3a shows a diagram of the relationship between online users of three above classes in the social network, under the influence of external shock effect $\epsilon(t)$ and growth effect $\eta(t)$. Here,

- β represents the rate of effective contacts between citizens in Infective and Susceptible classes. The higher the value of β is, the more the users will be "infected" by the Infective users.
- δ is the rate at which infected citizens lost interest in the topic and stop searching for it.

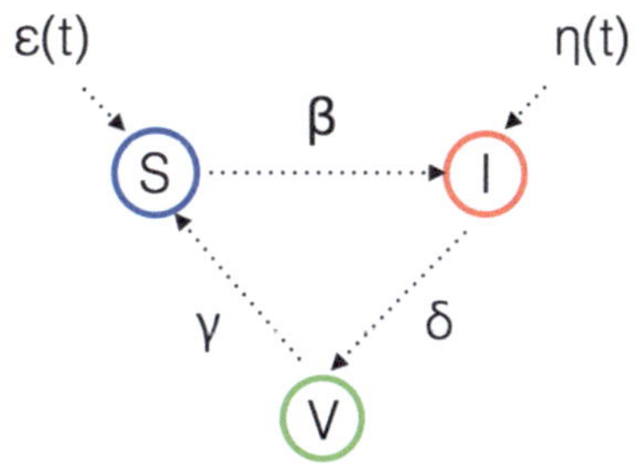

Fig. 2 *Δ*-SPOT diagrams: classes of population: susceptibles, infectives, and vigilants

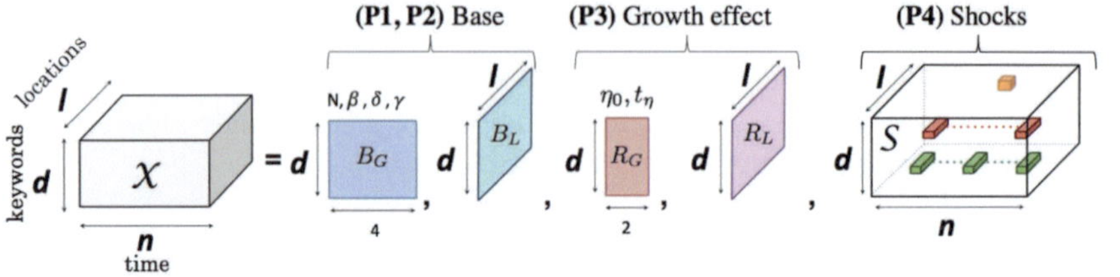

(a) Full parameter set (i.e., $\mathcal{F} = \{\mathbf{B_G}, \mathbf{B_L}, \mathbf{R_G}, \mathbf{R_L}, \mathcal{S}\}$)

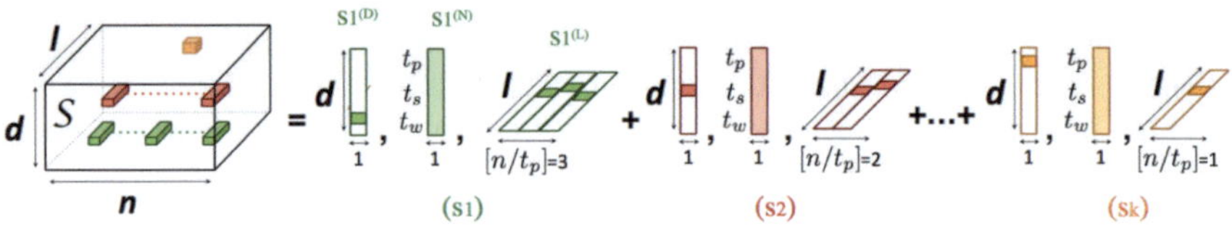

(b) External shock tensor (i.e., $\mathcal{S} = \{\mathbf{s}_1, \mathbf{s}_2, \ldots, \mathbf{s}_k\}$)

Fig. 3 **a** Δ-SPOT diagram, **b** Δ-SPOT structure: important properties extracted from tensor $\mathcal{X}$, also **c** external shock tensor $\mathcal{S}$ consists of a set of k components

- γ is the immunization loss probability for a change in status: being ready to search for the topic. When a new external event is about to happen, the value of γ increases as users become available to the news about it.

We also introduce two more parameter, $\epsilon(t)$ and $\eta(t)$, to represent the external shock effect and growth effect, respectively. The idea is that the number of the susceptible class $S(t)$ is the count of users available for infection, and if there is an external shock event at time-tick t, the infection becomes stronger than usual. Therefore, each infective pair would lead to a new infective citizen and will eventually cause a major spike. With respect to the growth effect, it starts at time t_η and make the number of infectives rise quickly to a new base.

Consequently, to handle all properties of online user activities, we introduce our model, namely, Δ-SPOT:

Model 1 (Δ- SPOT-single) *Our model can be described as the following equations*:

$$\begin{aligned} S(t+1) &= S(t) - \beta S(t)\epsilon(t)I(t)(1+\eta(t)) + \gamma V(t) \\ I(t+1) &= I(t) + \beta S(t)\epsilon(t)I(t)(1+\eta(t)) - \delta I(t) \\ V(t+1) &= V(t) + \delta I(t) - \gamma V(t) \end{aligned} \quad (1)$$

where the growth effect started at time t_η and $\eta(t)$ is defined as

$$\eta(t) = \begin{cases} 0 & (t < t_\eta) \\ \eta_0 & (t \geq t_\eta) \end{cases}$$

In addition, we introduce the temporal susceptible rate, $\epsilon(t)$, which is defined as follows:

$$\epsilon(t) = 1 + \sum_{i=1}^{k} f(t; \mathbf{s}_i)$$

$$f(t; \mathbf{s}) = \begin{cases} \epsilon_0 & (t_s + t_p\lceil t/t_p \rceil < t < t_s + t_p\lceil t/t_p \rceil + t_w) \\ 0 & (else) \end{cases}$$

where k is the number of shocks, and if $k = 0$, then $\epsilon(t) = 1$.

Each external shock consists of the following important properties: $\mathbf{s} = \{t_p, t_s, t_w, \epsilon_0\}$, i.e.,

- t_p is the periodicity of the event (if $t_p = \infty$, the event is noncyclic),
- t_s is the starting point of the event,
- t_w is the duration of the event, and
- ϵ_0 is the strength of the external shock.

3.3 *Δ*-SPOT—*with Multi-evolving Sequences*

So far we have seen how Δ-SPOT captures the dynamics of a single sequence. The next question is how can we apply Δ-SPOT to multiple time-evolving sequences in $\mathcal{X}$ and capture the individual behavior of d keywords in l locations/countries? We want to estimate the parameter set of Δ-SPOT, for each individual sequence in $\mathcal{X}$. We are interested in capturing global-level patterns, as well as local-level trends. So we thus propose "sharing" the global-level parameters for all l countries, to achieve effective modeling. Next, we want to extract the main trends and external patterns of co-evolving sequences $\mathcal{X} \in \mathbb{N}^{d \times l \times n}$, and make a good representation of $\mathcal{X}$. Figure 3b–c shows our modeling framework. Given a tensor $\mathcal{X}$, it extracts important patterns with respect to the following aspects, base properties of global and local trends $\mathbf{B_G}$, $\mathbf{B_L}$, population growth effect $\mathbf{R_G}$, $\mathbf{R_L}$, and external shock events $\mathcal{S}$. The first two are global- and local-level parameter sets, while the third also consists of both global and local-level parameter sets; however, the evaluating process is different.

Definition 1 (*Complete set of* Δ-SPOT) Let $\mathcal{F}$ be a complete set of parameters (namely, $\mathcal{F} = \{\mathbf{B_G}, \mathbf{B_L}, \mathbf{R_G}, \mathbf{R_L}, \mathcal{S}\}$) that describe the global/local patterns of the sequences in $\mathcal{X}$.

(P1) Base trends and global influence.

Definition 2 (*Base global matrix* $\mathbf{B_G}$ $(d \times 4)$) Let $\mathbf{B_G}$ be the set of global parameters of d keywords/queries, where $\{N_i, \beta_i, \delta_i, \gamma_i\}$ is the parameter set of the ith keyword.

We assume that the parameters are the same for all l countries. For example, the potential infection rate of each keyword (e.g., "Harry Potter") is the same for US and JP.

(P2) Area specificity. In the preceding discussion, we introduced the global-level matrices, which capture the common trends in $\mathcal{X}$. Next, we also want to analyze and explain location-specific patterns and trends in $\mathcal{X}$. For example, what is the difference of users' reaction for keyword "Ebola" between the U.S. (US) and Nepal (NP)? Our answer is that their behavior is similar, except for the "local sensitivity" of the sequence. Specifically, we share the parameters of the global-level matrices for all l countries, with one exception, N_{ij}, which describes the total popularity size of keyword i in the jth country. Specifically, we set the invariant, $N_{ij} = S_{ij}(t) + I_{ij}(t) + V_{ij}(t)$.

Definition 3 (*Base local matrix* $\mathbf{B_L}$ $(d \times l)$) Let $\mathbf{B_L}$ be a parameter set of the potential population of d keywords and l countries, i.e., $\mathbf{B_L} = \{b^{(L)}{}_{ij}\}_{i,j=1}^{d,l}$, where $b^{(L)}{}_{ij}$ is the potential population of <u>S</u>usceptibles of the ith keyword in the jth country.

This parameter corresponds to the fraction of individuals who are likely to be infected by the trend. For example, US has more users than NP, because they have more capacities for network connection.

(P3) Population growth effect. The growth effect appears due to the launch of new products and services that raise the interest of users, which should have the same starting time all over the world. Many sequences consist of the population growth effect, and Δ-SPOT should not detect and filter them as normal external shocks.

Definition 4 (*Growth global* $\mathbf{R_G}$ $(d \times 2)$) Let $\mathbf{R_G}$ be the set of global growth effect parameters of d keywords/queries, where $\{\eta_{0_i}, t_{\eta_i}\}$ is the parameter set of the ith keyword.

Definition 5 (*Growth local* $\mathbf{R_L}$ $(d \times l)$) Let $\mathbf{R_L}$ be a parameter set of the potential population of d keywords and l countries, i.e., $\mathbf{R_L} = \{r^{(L)}{}_{ij}\}_{i,j=1}^{d,l}$, where $r^{(L)}{}_{ij}$ is the population growth rate of the ith keyword in the jth country.

(P4) External shocks. We describe one external shock in terms of three aspects, *(keyword, country, time)*, for example, "Harry Potter, world-wide, Jul 15–21, 2007". To describe each external shock, we create a new parameter set, namely external shock tensor $\mathcal{S}$, which consists of a set of k external shocks, as described in Fig. 3c, i.e., $\mathcal{S} = \{\mathbf{s}_1, \mathbf{s}_2, \ldots, \mathbf{s}_k\}$. A single external shock $\mathbf{s}$ can be described as three components: $\mathbf{s} = \{\mathbf{s^{(D)}}, \mathbf{s^{(N)}}, \mathbf{s^{(L)}}\}$.

- Here, the $(d \times 1)$ size component $\mathbf{s^{(D)}}$ represents the external view for d keywords/queries;
- the (3×1) size component $\mathbf{s^{(N)}}$ describes the periodicity (t_p), the starting time (t_s), and the duration (t_w) of the external event;
- finally, the $(\lceil n/t_p \rceil \times l)$ size component $\mathbf{s^{(L)}}$ expresses the strength ϵ_0 of the external shock in l countries, where $\lceil n/t_p \rceil$ is the number of shocks belonging to one specified event.

Definition 6 (*External shock tensor* $\mathcal{S}$) Let $\mathcal{S}$ be a third-order tensor of k external shocks, i.e., $\mathcal{S} = \{\mathbf{s}_1, \mathbf{s}_2, \ldots, \mathbf{s}_k\}$, where the matrices show the parameters in terms of three components.

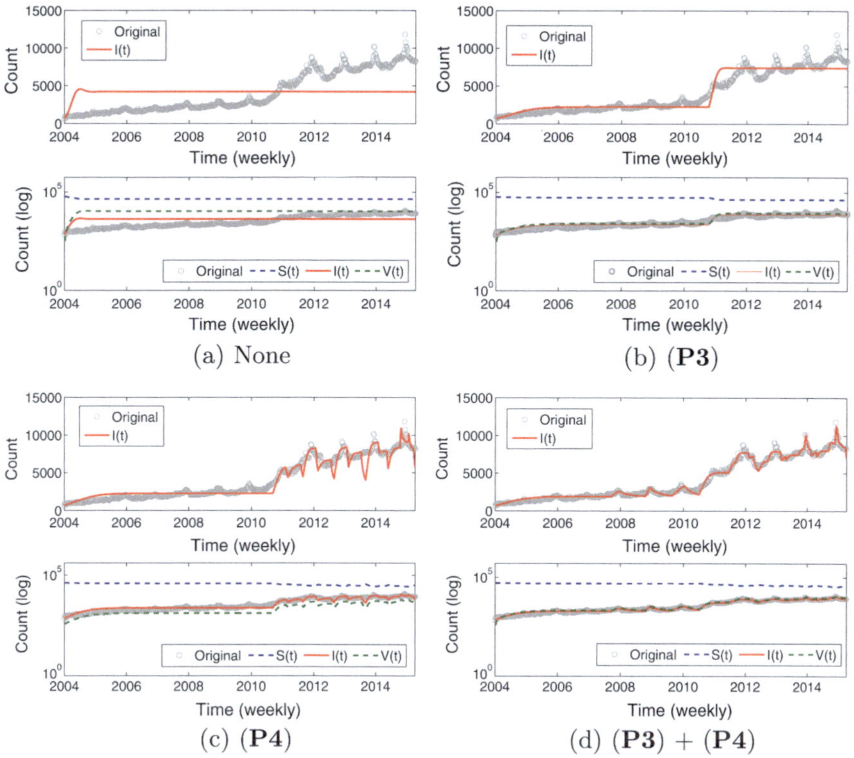

Fig. 4 Influence of combining growth effect and external shock effect: compared with the case of using **a** none of above, **b** only growth effect, **c** only external shock effect, and **d** the combination of both effects. Clearly, **d** fits the data very well

We propose to distinct between the growth effect (**P3**) and the external shock effect (**P4**). We use the growth effect to describe the rise of popularity size starting from a specified time-tick and last for long duration of time. On the other hand, the external shock effect analyzes the sudden rise of popularity size during a short period of time, which refers to a special event that happens in reality.

Figure 4 compares the fitting results (in both linear-scale and log-scale) of the keyword "Amazon", in four different cases to demonstrate the effectiveness of the growth effect (**P3**) and external shocks (**P4**). The result shows the benefit of treating the growth effect differently from external shock effect, as well as combining them to achieve good fitting results.[1]

[1]Here, the parameter values are $\beta = 5.014 \times 10^{-1}$, $\delta = 4.675 \times 10^{-1}$, $\gamma = 5.211 \times 10^{-1}$, $\eta_0 = 1.605 \times 10^{-1}$, $t_\eta = 343$ (the growth effect starts from time-tick 343).

4 Algorithm

In this section, we describe our fitting algorithm, which is carefully built to solve the following problem: Our goal is to extract the important patterns of online user activities from $\mathcal{X}$. More specifically, the problem that we want to solve is as follows:

Problem 1 **Given** a tensor $\mathcal{X}$ of $(keyword, country, time)$ triplets, **Find** a compact description that best summarizes $\mathcal{X}$, that is, $\mathcal{F} = \{\mathbf{B_G}, \mathbf{B_L}, \mathbf{R_G}, \mathbf{R_L}, \mathcal{S}\}$.

We want to find a good representation $\mathcal{F}$ to solve the problem. The essential questions are (a) How can we estimate the parameter set that best captures the dynamics and patterns in $\mathcal{X}$? (b) How can we find the optimal number of external shocks k? and (c) How can we treat an event to be cyclic or not?

4.1 Model Quality and Data Compression

Based on the minimum description length (MDL) principle, we propose an intuitive coding scheme. Here, it follows the assumption that the more we can compress data, the more we can detect its hidden patterns. In other words, we apply MDL to find an optimal representation $\mathcal{F}$.
Model description cost. The description complexity of model parameter set consists of two terms.

- First, the number of keywords d, locations l, and time-ticks n require $\log^*(d) + \log^*(l) + \log^*(n)$ bits.[2]
- Second, the model parameter set of the global base ($\mathbf{B_G}$), global growth effect ($\mathbf{R_G}$), and local base, growth effect ($\mathbf{B_L}, \mathbf{R_L}$), matrices require $d \times 4$, $d \times 2$, and $d \times l$ parameters, respectively, i.e., $Cost_M(\mathbf{B_G}) + Cost_M(\mathbf{R_G}) + Cost_M(\mathbf{B_L}) + Cost_M(\mathbf{R_L}) = c_F \cdot d(4 + 2 + l)$, where c_F is the floating point cost.[3]

Similarly, the model description cost of the external shock tensor $\mathcal{S} = \{\mathbf{s}_1, \mathbf{s}_2, \ldots, \mathbf{s}_k\}$ consists of the number of external shocks k ($\log^*(k)$ bits required), the shock-keyword vector $\mathbf{s}^{(\mathbf{D})}$ ($\log(d)$ bits required), the shock-time vector $\mathbf{s}^{(\mathbf{N})} = \{t_p, t_s, t_w\}$ ($3 \cdot \log(n)$ required), and the shock-location matrix $\mathbf{s}^{(\mathbf{L})}$ ($|\mathbf{s}^{(\mathbf{L})}| \cdot (\log(d) + \log(l) + \log(n) + c_F)$ required). Note that, for each shock $\mathbf{s}$, it requires $Cost_M(\mathbf{s}) = Cost_M(\mathbf{s}^{(\mathbf{D})}) + Cost_M(\mathbf{s}^{(\mathbf{N})}) + Cost_M(\mathbf{s}^{(\mathbf{L})})$. Consequently, the model cost of the external shock tensor $\mathcal{S} = \{\mathbf{s}_1, \cdots, \mathbf{s}_k\}$ is $Cost_M(\mathcal{S}) = \log^*(k) + \sum_{i=1}^{k} Cost_M(\mathbf{s}_i)$.
Data coding cost. Once we have decided the full parameter set $\mathcal{F}$, we can encode the data $\mathcal{X}$ with a given parameters $\mathcal{F}$:

[2]Here, $\log^*$ is the universal code length for integers, defined as $\log^*(x) \approx \log_2(x) + \log_2 \log_2(x) + \ldots$, where only the positive terms are included [21].

[3]We used 4×8 bits in our setting.

Algorithm 1 $\Delta - SPOT(\mathcal{X})$

1: **Input:** Tensor $\mathcal{X}$ $(d \times l \times n)$
2: **Output:** Full parameters, i.e., $\mathcal{F} = \{\mathbf{B_G}, \mathbf{B_L}, \mathbf{R_G}, \mathbf{R_L}, \mathcal{S}\}$
3: $\{\mathcal{F_G}\}$ =GLOBALFIT $(\mathcal{X})$; /* Parameter fitting for global-level sequences */
4: $\{\mathcal{F_L}\}$ =LOCALFIT $(\mathcal{X}, \mathcal{F_G})$; /* Parameter fitting for local-level sequences */
5: **return** $\mathcal{F} = \{\mathcal{F_G}, \mathcal{F_L}\}$;

$Cost_C(\mathcal{X}|\mathcal{F}) = \sum_{i,j,t=1}^{d,l,n} \log_2 p^{-1}_{Gauss(\mu,\sigma^2)}(x_{ij}(t) - I_{ij}(t))$, where $x_{ij}(t)$ is the elements in $\mathcal{X}$ and $I_{ij}(t)$ is the estimated count of infections (i.e., Model 1).[4]
Data compression equation. Consequently, the total code length for $\mathcal{X}$ with respect to a given parameter set $\mathcal{F}$ can be described in the following equation, which we want to minimize as our next goal:

$$Cost_T(\mathcal{X}; \mathcal{F}) = \log^*(d) + \log^*(l) + \log^*(n) + Cost_M(\mathbf{B_G}) + Cost_M(\mathbf{B_L}) + Cost_M(\mathbf{R_G}) + Cost_M(\mathbf{R_L}) + Cost_M(\mathcal{S}) + Cost_C(\mathcal{X}|\mathcal{F}) \quad (2)$$

4.2 *Multilayer Optimization*

Until now, we have shown how we can measure the goodness of the representation of $\mathcal{X}$, if we are given a candidate parameter set $\mathcal{F}$. Next, we explain how to find an optimal solution of the full parameter set: $\mathcal{F} = \{\mathbf{B_G}, \mathbf{B_L}, \mathbf{R_G}, \mathbf{R_L}, \mathcal{S}\}$. As described in Sect. 3.3, Δ-SPOT consists of multiple parameter sets, each of which explains either the local or global pattern of web search sequences in $\mathcal{X}$. In more details, the base and growth effect matrices $\mathbf{B_G}$ and $\mathbf{R_G}$ explain the global-level behavior of each time-series sequence, while the matrices $\mathbf{B_L}$ and $\mathbf{R_L}$ describe the local-level trends. Also, the extra tensor $\mathcal{S}$ consists of both global and local parameters. In order to estimate these model parameters, we propose a multilayer optimization algorithm, to search for the appropriate solution in terms of both global and local parameters. The idea is that we split parameter set $\mathcal{F}$ into two subsets, i.e., $\mathcal{F_G}$ and $\mathcal{F_L}$, each of which corresponds to a global/local-level parameter set, and try to fit the parameter sets separately. Our algorithm consists of the following two phases: GLOBALFIT and LOCALFIT.

- GLOBALFIT: find optimal global-level parameters for $\{\bar{\boldsymbol{x}}_i\}_{i=1}^{d}$, i.e., $\mathcal{F_G} = \{\mathbf{B_G}, \mathbf{R_G}, \mathcal{S}\}$
- LOCALFIT: find optimal local-level parameters: for $\{\boldsymbol{x}_{ij}\}_{i,j=1}^{d,l}$, i.e., $\mathcal{F_L} = \{\mathbf{B_L}, \mathbf{R_L}, \mathcal{S}\}$

[4]Here, μ and σ^2 are the mean and variance of the distance between the original and estimated values, and they need $2c_F$ bits, but we can eliminate them because they are constant values and independent of our modeling.

Algorithm 2 GLOBALFIT($\mathcal{X}$)

Input: Tensor $\mathcal{X}$
Output: Set of global-level parameters $\mathcal{F}_{\mathcal{G}}$ i.e., $\mathcal{F}_{\mathcal{G}} = \{\mathbf{B_G}, \mathbf{R_G}, \mathcal{S}\}$
for $i = 1 : d$ **do**
  Create $\bar{\boldsymbol{x}}_i$ from $\mathcal{X}$; /* Global sequence $\bar{\boldsymbol{x}}_i$ of i-th keyword */
  $\mathbf{s}_i = \emptyset$; /* Initialize external shocks for keyword i */
  while improving the cost **do**
    $\mathbf{b^{(G)}}_i = \arg\min_{\mathbf{b^{(G)}}'_i} \mathbf{Cost_C}(\bar{\mathbf{x}}_\mathrm{i} | \mathbf{b^{(G)}}'_\mathrm{i}, \mathbf{r^{(G)}}_\mathrm{i}, \mathbf{s}_\mathrm{i})$; /* Base */
    $\mathbf{r^{(G)}}_i = \arg\min_{\mathbf{r^{(G)}}'_i} \mathbf{Cost_C}(\bar{\mathbf{x}}_\mathrm{i} | \mathbf{b^{(G)}}_\mathrm{i}, \mathbf{r^{(G)}}'_\mathrm{i}, \mathbf{s}_\mathrm{i})$; /* Growth */
    $\mathbf{s}_i = \emptyset$; /* Initialize values */
    /* Find external shocks for keyword i */
    while improving the cost **do**
      $\mathbf{s} = \arg\min_{\mathbf{s}'} Cost_C(\bar{\boldsymbol{x}}_i | \mathbf{b^{(G)}}_\mathrm{i}, \mathbf{r^{(G)}}_\mathrm{i}, \{\mathbf{s}_\mathrm{i} \cup \mathbf{s}'\})$; $\mathbf{s}_\mathrm{i} = \mathbf{s}_\mathrm{i} \cup \mathbf{s}$;
    end while
  end while
  $\mathbf{B_G} = \mathbf{B_G} \cup \mathbf{b^{(G)}}_\mathrm{i}$; $\mathbf{R_G} = \mathbf{R_G} \cup \mathbf{r^{(G)}}_\mathrm{i}$; $\mathcal{S} = \mathcal{S} \cup \mathbf{s}_\mathrm{i}$; /* Update parameter set of i-th keyword */
end for
return $\mathcal{F}_{\mathcal{G}} = \{\mathbf{B_G}, \mathbf{R_G}, \mathcal{S}\}$;

Here, the global-level sequence of the ith keyword: $\bar{\boldsymbol{x}}_i$ can be described as the sum of d local sequences, i.e., $\bar{\boldsymbol{x}}_i(t) = \sum_{j=1}^{l} \boldsymbol{x}_{ij}(t)$. Algorithm 1 shows an overview of Δ-SPOT to find the full parameter set given a tensor $\mathcal{X}$.

Global-level parameter fitting. Given a tensor $\mathcal{X}$, our subgoal is to find the optimal global parameter set: $\mathcal{F}_{\mathcal{G}}$, to minimize the cost function (i.e., Eq. 2). So how can we fit the parameters as well as simultaneously estimate the appropriate number of external shocks? To find a good basic parameter set for $\mathcal{X}$, we have to filter out the external shocks. Also, a good external shock filter requires a well-estimated model. We escape this circular dependency by applying an iterative method that employs external shocks detection and filtering, and model fitting in an alternating way until the cost function reaches a minimum value. We keep detecting new external shocks as long as the total cost function becomes lower, and the number of detected external shocks that minimizes the cost is considered to be optimal.

As shown in Algorithm 2, we provide a detailed algorithm of the global-level fitting. Given a tensor $\mathcal{X}$, it creates a set of d global sequences: $\{\bar{\boldsymbol{x}}_i\}_{i=1}^{d}$. The goal is to fit the global-level parameter set, as well as find the appropriate number of external shocks. We apply the *Levenberg–Marquardt* (*LM*) [6] algorithm to minimize the cost function. Note that the extra tensor $\mathcal{S}$ consists of k entries $\{\mathbf{s}_1, \mathbf{s}_2, \ldots, \mathbf{s}_k\}$, and Algorithm 2 can find only the global-level entry, which consists of $(keyword, time)$. We will introduce the local-level parameter fitting algorithm in Algorithm 3, to describe how the local-level entries can be computed. Also, the cost function (2) includes the cost of local-level parameters such as $\mathbf{B_L}$ and $\mathbf{R_L}$ but these terms are independent of the global model fitting. Hence, we can simply consider them to be constant.

Algorithm 3 LOCALFIT($\mathcal{X}$, $\mathbf{B_G}$, $\mathbf{R_G}$, $\mathcal{S}$)

Input: (a) Tensor $\mathcal{X}$, (b) global-level parameter set $\mathcal{F_G}$
Output: Set of local-level parameters, i.e., $\mathcal{F_L}$
while improving the cost **do**
  /* For each local sequence $\boldsymbol{x}_{ij}$ of i-th keyword in j-th country */
  for $i = 1 : n$ **do**
    for $j = 1 : l$ **do**
      $b^{(L)}{}_{ij} = \arg\min_{b^{(L)'}_{ij}} Cost_C(\boldsymbol{x}_{ij}|\mathbf{B_G}, \mathbf{R_G}, b^{(L)'}_{ij}, \mathcal{S})$;
      $r^{(L)}{}_{ij} = \arg\min_{r^{(L)'}_{ij}} Cost_C(\boldsymbol{x}_{ij}|\mathbf{B_G}, \mathbf{R_G}, r^{(L)'}_{ij}, \mathcal{S})$;
    end for
  end for
  for each external shock **s** in $\mathcal{S}$ **do**
    Update **s** to minimize the cost /* Local participation rate */
  end for
end while
return $\mathcal{F_L} = \{\mathbf{B_L}, \mathbf{R_L}, \mathcal{S}\}$;

Local-level parameter fitting. Given a set of $d \times l$ local-level sequences, $\{\boldsymbol{x}_{ij}\}_{i,j=1}^{d,l} \in \mathcal{X}$, and a set of global-level parameters, $\mathcal{F_G}$, our next goal is to fit the individual parameters of each keyword in each country, that is, $\mathcal{F_L} = \{\mathbf{B_L}, \mathcal{S}\}$. We propose an iterative optimization algorithm (see Algorithm 3). Our algorithm searches for the optimal solution with respect to (a) the base local matrix $\mathbf{B_L}$ and (b) the local-level external shocks $\mathcal{S}$, so that the total coding cost is minimized. Consequently, the computation time of Δ-SPOT is $O(dln)$.

Lemma 1 *The computation time of* Δ-SPOT *is* $O(dln)$.

Proof The algorithm requires $O(dln)$ time, to create the global-level sequences from $\mathcal{X}$. For global-level parameter fitting, it needs $O(\#iter \cdot k \cdot dn)$ time, where $\#iter$ is the number of iterations, k shows the number of external shocks. For the local-level parameter fitting, it requires $O(\#iter \cdot k \cdot dln)$ time to fit the parameters. Note that $\#iter$, k are small constant values that are negligible. Thus, the complexity is $O(dln)$.

5 Experiments

In this section, we demonstrate the effectiveness of Δ-SPOT with real dataset. The experiments were designed to answer the following questions:

Q1 *Sensemaking*: Can our method help us understand the given input online activities?
Q2 *Accuracy*: How well does our method match the data?
Q3 *Scalability*: How does our method scale in terms of computational time?

Dataset description. We performed experiments on three real datasets:

1. *GoogleTrends*: This dataset consists of the volume of searches for queries (i.e., keywords) in various topics (i.e., events, celebrities, movies, etc.) on Google[5] from January 2004 to January 2015 (in weekly basis), collected in 232 countries. Each query represents the search volumes that are related to keywords over time (in weekly basis).
2. *Twitter*: We used more than seven million Twitter[6] posts covering an 8-month period from June 2011 to January 2012. We selected the 10,000 most frequently used hashtags, in a 1-week window around their peak volume, with 100,000 users that mentioned these items most frequently. For each hashtag, we used a week window around the peak volume.
3. *MemeTracker*: This dataset covers 3 months of blog activity from August 1 to October 31, 2008,[7] It contains more than 70,000 short quoted textual phrases ("memes"), each of which consists of the number of mentions over time. We choose 1000 phrases in blogs with the highest volume in a 7-day window around their peak volume.

5.1 Sensemaking

In this experiment, we demonstrate how effective Δ-SPOT can be in terms of data fitting, external events detection, and other important properties. Our objective is to fit the popularity size of the Infective class. We demonstrate the global fitting results of several sequences collected from three above datasets:

1. Figure 5 shows the results of model fitting on four trending keywords/queries in various categories.
2. Figure 6 shows the results of two popular hashtags "#apple" and "#backtoschool".
3. Figure 7 shows the results of two phrases ("meme")[8] from the *MemeTracker* dataset.

In all above figures, we show the original sequences (i.e., black dots) and estimated sequences: $I(t)$ (i.e., red line) in linear–linear scales. Also, we made several important observations, which correspond to the properties mentioned above.

- **(P1) Base trends and global influence**. As shown in Fig. 5, our proposed model successfully captures long-range nonlinear dynamics of user activities, as well

[5]http://www.google.com/insights/search/.

[6]http://twitter.com/.

[7]http://memetracker.org/.

[8]Meme#3:"yes we can yes we can"
Meme#16:"joe satriani is a great musician but he did not write or have any influence on the song viva la vida we respectfully ask him to accept our assurances of this and wish him well with all future endeavours".

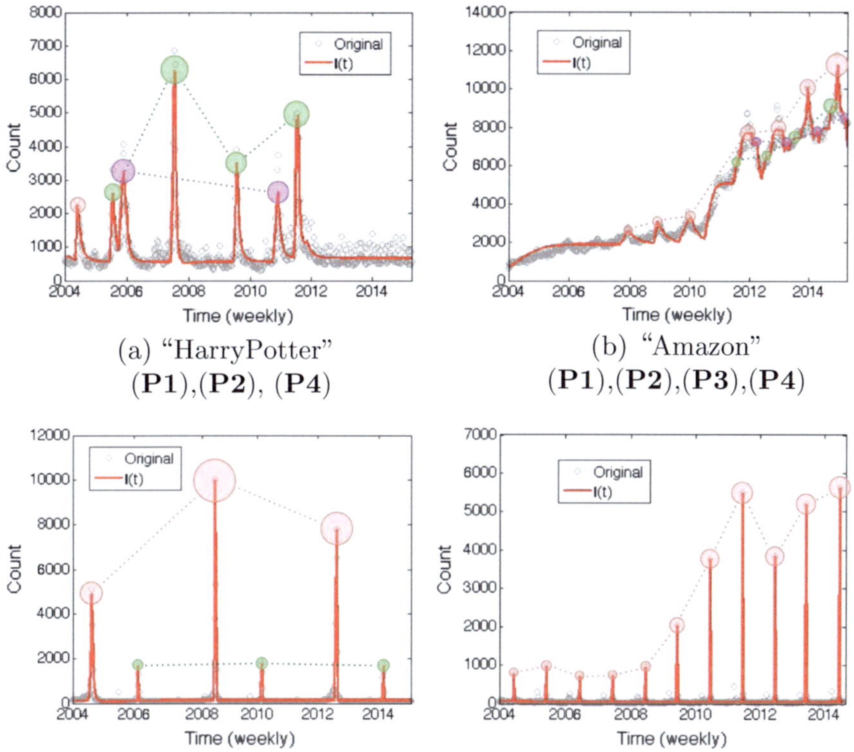

(a) "HarryPotter" (P1),(P2), (P4)

(b) "Amazon" (P1),(P2),(P3),(P4)

(c) "Olympics" (P1),(P2),(P4)

(d) "Grammy" (P1),(P2),(P4)

Fig. 5 Global fitting results for four queries in GoogleTrends dataset of different topics (celebrities, sporting events, awards, movies, etc.)

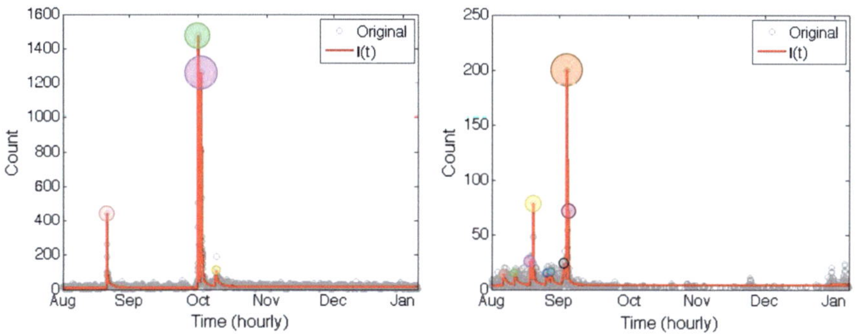

(a) "#apple" (P1),(P2),(P4)

(b) "#backtoschool" (P1),(P2),(P4)

Fig. 6 Global fitting results for two hashtags in Twitter dataset

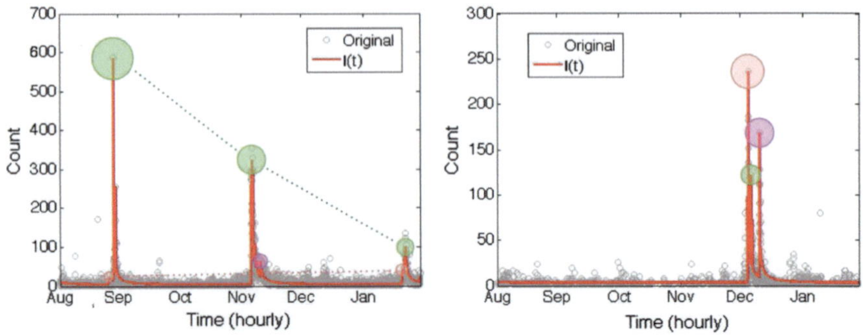

(a) Meme#3
(P1),**(P2)**,**(P4)**

(b) Meme#16
(P1),**(P2)**,**(P4)**

Fig. 7 Global fitting results for two memes in MemeTracker dataset

as fit the data sequences in high accuracy. We also demonstrate the effectiveness of Δ-SPOT in linear-scale and log-scale fitting. Figure 8 shows the fitting result for keyword "Ice Cream" in *GoogleTrends* dataset, in which the upper and lower figure display the linear-scale and log-scale fitting result, respectively. Δ-SPOT proves to well fit the sequence in both scales, and it also effectively describes the basic dynamic of the time series.

- **(P2) Area specificity**. Figure 9a shows the local fitting results for keyword "Ebola" of *GoogleTrends* dataset; in which we detected some countries (AU, RU, GB, US, and JP) that behave similar to the global trend (i.e., the world reaction to the burst of Ebola Virus in 2014, shown in green circles). Besides, we can also detect several outliers from the countries which have less capacities of network connection (LA, NP, CG).
- **(P3) Population growth effect**. In Fig. 5b, we detected the growth effect in keyword "Amazon", which starts from time-tick 343. In fact, this relates to the development of online service since 2011, which leads to the quick rise in the number of users who search for the Amazon services, including online shopping, media downloads, and other cloud infrastructure services. This result proves that our model can accurately capture the population growth effect to describe the true dynamic of the time series, and also distinguish its behavior from the external shock effect.
- **(P4) External shock events**. Δ-SPOT can capture important external events relating to the keywords, including the cyclic events. Furthermore, Δ-SPOT can capture the cyclic external events of different periodicities and durations, as shown in Fig. 5. For better visualization, we mark the cyclic external shocks by the same color circles. All external shocks that belonged to a cyclic event are marked by the same color, and also linked by dot lines to emphasize the cyclic pattern. Δ-SPOT guarantees not to miss any important external events that happen during the sequences.

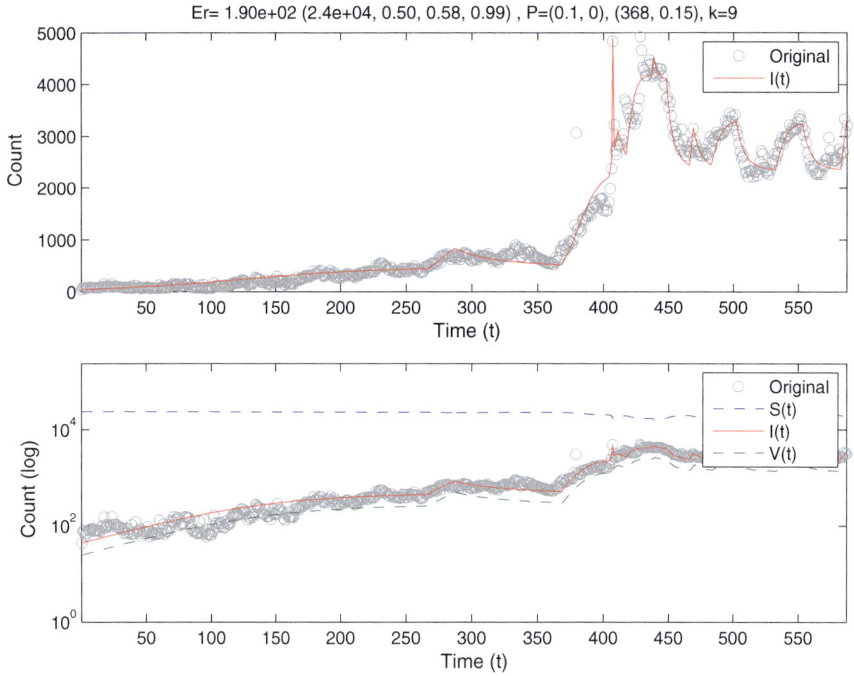

Fig. 8 Linear-scale and log-scale fitting for keyword "Ice Cream" in *GoogleTrends* dataset: our method can fit the sequences in both scales with high accuracy

5.2 *Accuracy*

Next, we discuss the quality of Δ-SPOT in terms of fitting accuracy. We used the fitting result for keyword "Amazon" of *GoogleTrends* dataset, and compared Δ-SPOT with the standard *SIRS* model, *SKIPS* [22], and multi-time-evolving sequences mining model, *FUNNEL* [16]. Figure10a shows the root-mean-square error (RMSE) between the original and estimated counts of the global sequences $\{\bar{x}_i(t)\}_{i,t}^{d,n}$. Similarly, Fig. 10b shows the results of the local counts $\{x_{ij}(t)\}_{i,j,t}^{d,l,n}$, (i.e., each keyword in each country, at each time-tick). A lower value indicates a better fitting accuracy. Note that the *SIRS* model cannot capture seasonal dynamics, *SKIPS* has the ability to capture periodic patterns, while *FUNNEL* can capture external shocks. Moreover, they are not intended to detect growth effect. As shown in the figures, the *SIRS* model and *SKIPS* failed to capture the complicated patterns of data sequences, *FUNNEL* cannot detect cyclic external events, while our method achieved those properties with high fitting accuracy. We also evaluated the accuracy of Δ- STREAM in terms of global/local fitting. Δ- STREAM still provides better fitting accuracy compared to other methods.

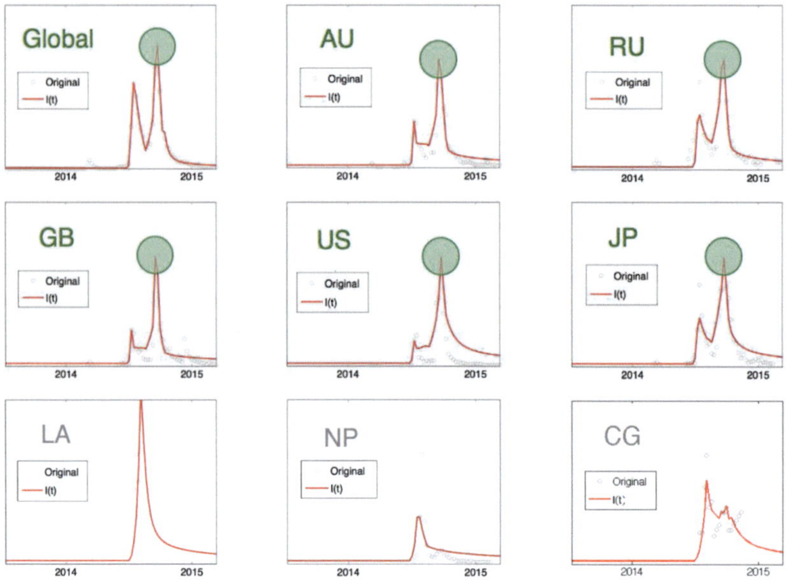

(a) Original/fitted sequences for "Ebola"

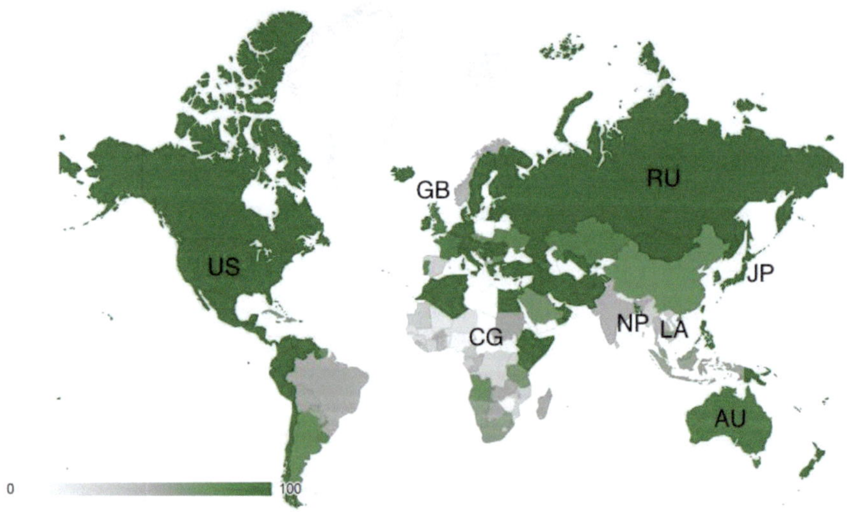

(b) World-wide reaction

Fig. 9 Local fitting power of Δ-SPOT for the keyword "Ebola" which refers to the Ebola Virus bursting in 2014 (shown in green circles). **a** It can capture the local similar behaviors in Australia (AU), Russia (RU), the U.K. (GB), the U.S. (US), and Japan (JP). It can also capture local outliers in Laos (LA), Nepal (NP), and DR Congo (CG), in comparison to the global trend. And we have a clearer observation in (**b**) the world map of user reaction to the disease burst in 2014

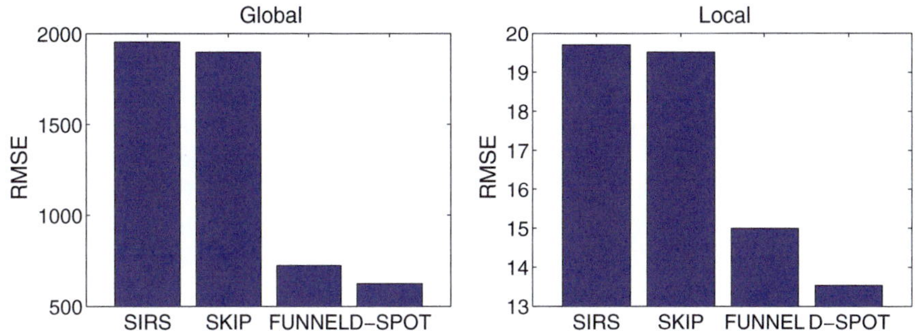

(a) Global fitting (b) Local fitting

Fig. 10 Fitting accuracy for the **a** global sequences: $\{\bar{x}_i(t)\}_{i,t}^{d,n}$ and **b** local sequences: $\{x_{ij}(t)\}_{i,j,t}^{d,l,n}$. Δ-SPOT consistently outperforms the previous models w.r.t. the RMSE between real and estimated values (lower is better)

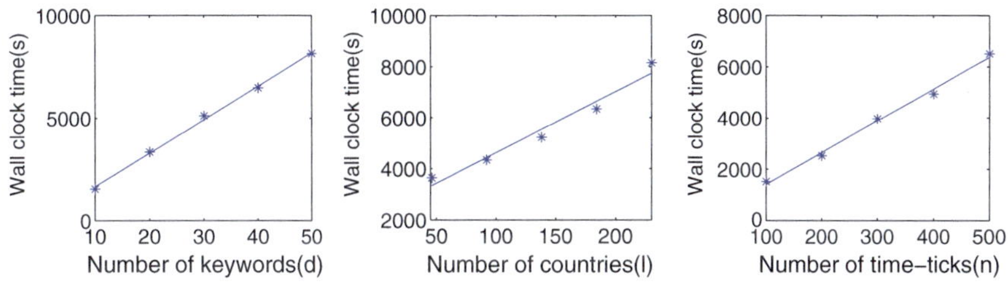

(a) Keywords (d) (b) Countries (l) (c) Duration (n)

Fig. 11 Δ-SPOT scales linearly: Wall clock time versus data size $(d \times l \times n)$. Our method is linear on the data size

5.3 Scalability

We made the evaluation of the scalability of Δ-SPOT and verified the complexity of our method, which we discussed in Sect. 4. As shown in Fig. 11, Δ-SPOT is linear with respect to (a) keywords d, (b) countries l, and (c) duration n. More importantly, our proposed online streaming method, Δ- STREAM achieves a dramatic reduction in computational time, thanks to our coding scheme that reduces the processing time of external shocks detection and parameter set updating. Especially, with respect to the duration of time, it requires constant, i.e., it does not depend on n.

6 $\boldsymbol{\Delta}$-SPOT at Work

As we described in the previous section, Δ-SPOT is capable of analyzing online activities of various categories. In this section, we discuss the most important and challenging task of Δ-SPOT, namely, forecasting the future dynamics of co-evolving activities. The goal here is to predict the future dynamics of the online activities, including the basic pattern and the external events of the sequences. As described in Sect. 4, Δ-SPOT can detect the exact periodicity of the cyclic events. Given the external shock tensor $\mathcal{S}$, Δ-SPOT automatically generates the next shocks of the cyclic events in terms of the time and duration, respectively. Here, we use the regression function to estimate the strength of those shocks, given the strength of the previous shocks. Figure 12 shows results of our forecasting for keyword "Grammy". We trained the model parameters by using the 400 time-ticks of the sequence (solid black lines in the figures), and then did forecasting on the following years (solid red lines). The result shows that Δ-SPOT can predict the time-tick, the duration, and the relative strength of incoming external events, which refer to the annual Grammy Awards, held every February. We also make a comparison between Δ-SPOT and other competing methods, including the auto-regressive (AR) model, and TBATS model.

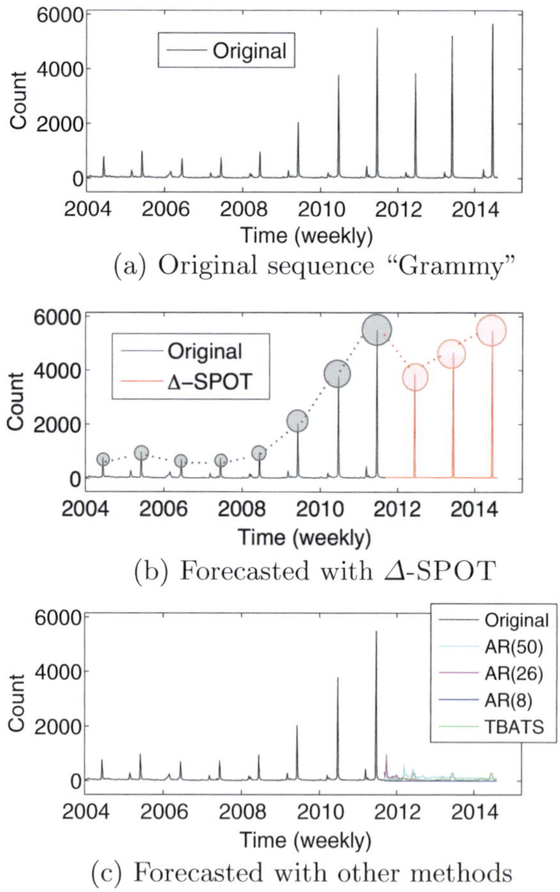

Fig. 12 Forecasting result: we train the model parameters using first 400 time-ticks of the sequences and do forecasting the remaining part

For AR, We applied several regression coefficients: $r = 8, 26, 50$. In Fig. 12a, b, c, we show the original sequences, the forecast results of Δ-SPOT, and AR with TBATS, respectively. It is clearly shown that our method can predict the next three external shocks, which refer to the next three Grammy Awards. On the other side, AR and TBATS fail to capture the accurate future dynamics, and also cannot detect the future external events.

7 Conclusion

In this paper, we presented Δ-SPOT, an intuitive model for mining large-scale time-evolving online activities. Through the extensive experiments on real datasets, Δ-SPOT demonstrates all the following desirable properties:

1. It is **effective**: it can detect important hidden events that match the reality, as well as achieves high accuracy.
2. It is **automatic**: it requires no training set and no domain expertise.
3. It is **scalable**: Δ-SPOT is linear to the data size (i.e., $O(dln)$).
4. It is **practical**: Δ-SPOT can undertake long-range forecasting and outperforms existing methods.

Acknowledgements This work was partially supported by JSPS KAKENHI Grant-in-Aid for Scientific Research Number JP15H02705, JP17H04681, JP16K12430, PRESTO JST, the MIC/SCOPE #162110003, and the ICT infrastructure establishment for clinical and medical research from Japan Agency for Medical Research and development, AMED.

References

1. H. Choi and H. R. Varian. Predicting the present with google trends. *The Economic Record*, 88(s1):2–9, 2012.
2. J. Ginsberg, M. Mohebbi, R. Patel, L. Brammer, M. Smolinski, and L. Brilliant. Detecting influenza epidemics using search engine query data. *Nature*, 457:1012–1014, 2009.
3. S. Goel, J. Hofman, S. Lahaie, D. Pennock, and D. Watts. Predicting consumer behavior with web search. *PNAS*, 2010.
4. D. Gruhl, R. Guha, R. Kumar, J. Novak, and A. Tomkins. The predictive power of online chatter. In *KDD*, pages 78–87, 2005.
5. A. Jain, E. Y. Chang, and Y.-F. Wang. Adaptive stream resource management using kalman filters. In *SIGMOD*, pages 11–22, 2004.
6. K. Levenberg. A method for the solution of certain non-linear problems in least squares. *Quarterly Journal of Applied Mathmatics*, II(2):164–168, 1944.
7. L. Li, C.-J. M. Liang, J. Liu, S. Nath, A. Terzis, and C. Faloutsos. Thermocast: A cyber-physical forecasting model for data centers. In *KDD*, 2011.
8. L. Li, J. McCann, N. Pollard, and C. Faloutsos. Dynammo: Mining and summarization of coevolving sequences with missing values. In *KDD*, 2009.
9. L. Li, B. A. Prakash, and C. Faloutsos. Parsimonious linear fingerprinting for time series. *PVLDB*, 3(1):385–396, 2010.

10. A. M. D. Livera, R. J. Hyndman, and R. D. Snyder. Forecasting time series with complex seasonal patterns using exponential smoothing. *Journal of the American Statistical Association*, 106(496):1513–1527, 2011.
11. Y. Matsubara, Y. Sakurai, and C. Faloutsos. Autoplait: automatic mining of co-evolving time sequences. In *SIGMOD*, pages 193–204, 2014.
12. Y. Matsubara, Y. Sakurai, and C. Faloutsos. The web as a jungle: Non-linear dynamical systems for co-evolving online activities. In *WWW*, 2015.
13. Y. Matsubara, Y. Sakurai, and C. Faloutsos. Non-linear mining of competing local activities. In *WWW*, 2016.
14. Y. Matsubara, Y. Sakurai, C. Faloutsos, T. Iwata, and M. Yoshikawa. Fast mining and forecasting of complex time-stamped events. In *KDD*, pages 271–279, 2012.
15. Y. Matsubara, Y. Sakurai, B. A. Prakash, L. Li, and C. Faloutsos. Rise and fall patterns of information diffusion: model and implications. In *KDD*, pages 6–14, 2012.
16. Y. Matsubara, Y. Sakurai, W. G. van Panhuis, and C. Faloutsos. FUNNEL: automatic mining of spatially coevolving epidemics. In *KDD*, pages 105–114, 2014.
17. S. Papadimitriou, A. Brockwell, and C. Faloutsos. Adaptive, hands-off stream mining. In *VLDB*, pages 560–571, 2003.
18. B. A. Prakash, A. Beutel, R. Rosenfeld, and C. Faloutsos. Winner takes all: competing viruses or ideas on fair-play networks. In *WWW*, pages 1037–1046, 2012.
19. T. Preis, H. S. Moat, and H. E. Stanley. Quantifying trading behavior in financial markets using google trends. *Sci. Rep.*, 3, 04 2013.
20. T. Rakthanmanon, B. J. L. Campana, A. Mueen, G. E. A. P. A. Batista, M. B. Westover, Q. Zhu, J. Zakaria, and E. J. Keogh. Searching and mining trillions of time series subsequences under dynamic time warping. In *KDD*, pages 262–270, 2012.
21. J. Rissanen. A Universal Prior for Integers and Estimation by Minimum Description Length. *Ann. of Statist.*, 11(2):416–431, 1983.
22. L. Stone, R. Olinky, and A. Huppert. Seasonal dynamics of recurrent epidemics. *Nature*, 446:533–536, 2007.
23. Y. Tao, C. Faloutsos, D. Papadias, and B. Liu. Prediction and indexing of moving objects with unknown motion patterns. In *SIGMOD*, pages 611–622, 2004.

PEGASEF: A Provenance-Based Big Data Service Framework for Efficient Simulation Execution on Shared Computing Clusters

Young-Kyoon Suh, Ki Yong Lee and Nakhoon Baek

Abstract Over the past years high-performance computing (HPC) simulation programs have been aggressively employed to solve complex problems in a variety of computational science and engineering disciplines. As those programs are shared in an online platform, many users can easily run their simulations on the platform as long as they are connected on the web. However, repetitive simulations from users have charged a significant burden on the platform's limited computing and storage resources. To address the concern of inefficiency in simulation execution, we propose a big data service framework based on past simulation records. Such records are called *provenances*, which capture various properties in simulation. By utilizing the provenances, the platform can perform more efficient simulations via duplicate elimination and assist users with enhanced simulation service such as result prediction, execution-time estimation, and input-parameter clustering.

Keywords HPC · Efficient simulation · Provenance · Big data · Service framework · Duplicate elimination

---

Y.-K. Suh · N. Baek
School of Computer Science and Engineering, Kyungpook National University, Daegu 41566, Republic of Korea
e-mail: yksuh@knu.ac.kr

N. Baek
e-mail: oceancru@gmail.com

K. Y. Lee (✉)
Division of Computer Science, Sookmyoung Women's University, Seoul 04310, Republic of Korea
e-mail: kiyonglee@sookmyung.ac.kr

W. Lee and C. K. Leung (eds.), *Big Data Applications and Services 2017*,
Advances in Intelligent Systems and Computing 770,
https://doi.org/10.1007/978-981-13-0695-2_17

1 Introduction

Over the past years, computer simulation programs (tools) have been aggressively employed to solve complex problems in a variety of computational science and engineering (CSE) disciplines. Accordingly, if numerous users simultaneously have access to and submit simulation requests to a platform, then the platform transparently conducts requested simulations and returns the results to the users. While supporting a number of simulation users, such a platform makes full use of high-performance computing (HPC) clusters for seamlessly performing compute-bound simulations and high-end storage resources for storing a huge volume of results yielded by the simulations. Thanks to this cutting-edge Information Technology (IT) infrastructure, CSE users can run their simulations and obtain the results with little constraints.

One of the most popular online simulation platforms is EDISON [1, 2]. This platform was devised to support several CSE communities to carry out research and education by sharing simulation tools and running simulations on the tools. For those who are not quite familiar with command-line interface (CLI), the platform provides users with an easy-to-use interface to submit a simulation job and view the job's results as illustrated in Fig. 1. In addition, the completed simulation results can be freely downloadable at all times (as long as the platform is not in the period of platform maintenance). Hence, this platform has been a great playground for the selected CSE communities that desire to run their simulations on large-scale computing clusters and educate students with the shared simulation tools.

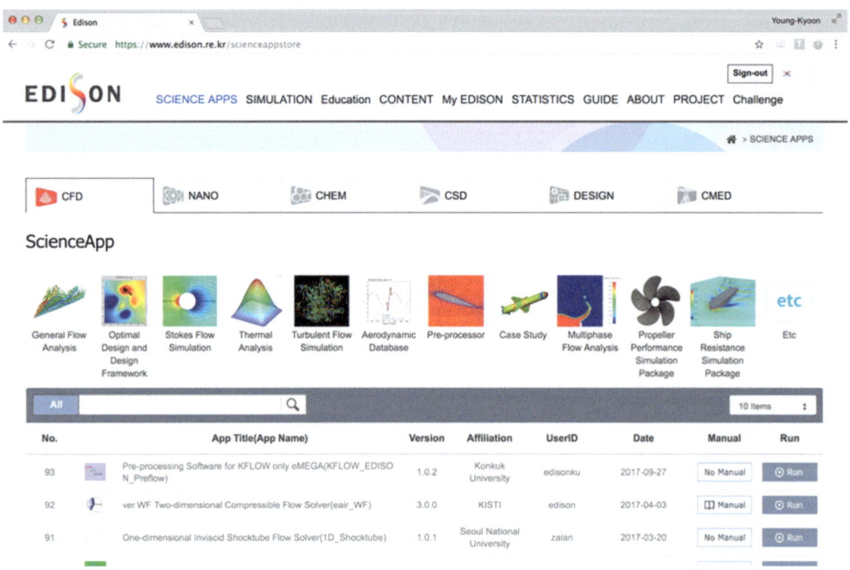

Fig. 1 The EDISON portal site

Despite the platform's versatility, one of the critical problems it faces is that different users may submit with *identical* input parameters the *same* computational jobs over and over again, thereby incurring a significant overhead. For instance, an entire class consisting of tens of students may choose the same simulation tool and run the same simulations without altering default input parameters at all. If that is the case, a great number of the identical computing jobs get submitted to the clusters, resulting in inefficient utilization of limited resources. Thus, other users, who can also have simultaneous accesses to the platform, may not be satisfied with the service due to the slowdown incurred by such duplicate simulations.

In fact, for completed simulations, a variety of information associated with the simulations can be recorded. Suppose that a certain user requests a simulation with specific input parameters at a certain timestamp. Once the simulation is completed, we know (i) *what user* initiated the simulation, (ii) *when* to execute the simulation, (iii) *what input parameters* or *files* were provided to that simulation, (iv) *how* the simulation ended—*success* or *failure*, and (v) *what output files* were produced if it succeeded. Besides, there would be more information that can be collected. This information is called *provenance* [3, 4].

In spite of its importance, little attention has been paid to using simulation provenance for efficient and intelligent simulation service for users. By making aggressive use of provenance, we can expect two benefits. One is to allow a platform to be able to reduce workloads through duplicate elimination of requested simulations. *Efficiency*, thus, is achieved. Another is to assist users to run their simulations, with being more informed from the "big" provenance data agglomerated by other users' past simulations. Hence, *intelligence* is achieved as well.

In this paper, we propose a novel big data service framework for efficiently executing HPC simulations by exploiting big provenance data. This framework provides several functionalities, such as (i) seamlessly collecting simulation provenances from the platform, (ii) retrieving provenances, (iii) estimating simulation results based on refined provenances stored in a repository, and (iv) provenance data mining such as execution-time estimation and input parameter set clustering. The framework is expected to significantly boost the performance of the online simulation platform, thereby achieving efficiency and intelligence in simulation.

The rest of this paper is organized in the following. The following section discusses some related work and provides several distinctions from our work. Then we propose and elaborate on our framework. Finally, we conclude our paper by summarizing our contributions.

2 Related Work

Recently, there has been a rich body of utilizing provenances associated with various kinds of scientific data, such as simulation output, experiments, measurement, and articles. It is useful to verify the validity of respective data if their provenance information is supplied together. Also, it is easy to reproduce the output through the provided provenance. To this end, many provenance-based systems have been proposed and actually harnessed.

Online simulation service platforms. HUBzero [5], which was created by Purdue University, aims at developing, retrieving, and running simulation tools online. Like the EDISON platform, its simulations are carried out on HPC infrastructure, and the completed results are stored in storage resources. nanoHUB [6] was derived from the HUBzero platform. This platform focuses on the nanotechnology research and academic communities in which a variety of simulation tools are shared and harnessed to run simulations. There are other types of simulation platforms such as DataHUB [7], SimulationHub [8], and WebMO [9]. But none of these platforms considers utilizing provenances for efficient and intelligent simulation for users.

Provenance-based scientific simulation platforms. myGrid [10] is a platform to perform simulations represented by a scientific workflow, and it is widely used in biology and bioinformatics. This platform can collect provenances from a completed workflow, but the collected provenances are used only for information browsing. Taverna [11] is another workflow system used in the myGrid project. It adds automated provenance collection while executing a workflow. But it does not seem to take into consideration more efficient workflow execution based on the captured provenances. Kepler [12] is a scientific workflow system and provides a feature to design a workflow via graphical user interface (GUI). This system adds a provenance framework in which it is possible to keep track of how a given workflow generates data during its execution. Although it may be feasible to re-execute the workflow based on the collected provenance, it does not provide advanced provenance-driven services such as execution-time estimation, input parameter clustering, and duplicate elimination.

Big data based provenance platforms. Reduce And Map Provenance (RAMP) [13] is a novel system to extend Hadoop to keep track of a workflow comprised of MapReduce jobs. In RAMP, it is possible to automatically collect detailed provenance on a certain workflow while being transparent to users and Hadoop. HadoopProv [14] is similar to RAMP except providing a feature to perform analytics on collected provenance records. Pig Lipstick [15] is a tracking system for a workflow executed on Apache Pig [16]. Users can visualize and monitor the flow of data involving the workflow. Again, none of these systems seem to provide any further intelligent service beyond provenance collection.

Cloud-based provenance platforms. CloudProv [17] is a framework to model, integrate, and monitor provenances on real-time data in a cloud environment. It provides public APIs for applications to share and integrate provenance data. As a cloud-based computing platform, e-Science Central [18] provides Software-as-a-service (Saas) and Platform-as-a-Service (PaaS) for scientific data management, analysis, and collaboration. Computational scientists can upload data, execute a workflow, and share the results on a cloud. e-Science Central provides collecting and visualizing workflow provenance. These cloud-based platforms are also interested in recording provenances for scientific workflow, but again those do not utilize the provenances for efficient execution of workflow and advanced mining service such as execution-time estimation.

3 PEGASEF: A Provenance-Driven Big Data Service Framework

In this section, we propose our framework, named PEGASEF (ProvEnance-driven biG dAta SErvice Framework), for efficient execution and advanced user-service of HPC simulation.

Figure 2 shows the PEGASEF framework, consisting of four components. The leftmost two components represent an online simulation platform such as EDISON. Note that our proposed framework is not solely applicable to the EDISON platform, and it can collaborate with any other simulation platform as long as collected provenances can be injected into the platform. We now elaborate on the framework.

Simulation Provenance Big Data Repository: This component is a permanent repository for storing simulation provenances collected on the platform.

Once a requested simulation is successfully finished, the platform collects a variety of provenance information about (1) *which simulation tool* and (2) *what version of the tool* was used, (3) *what input parameters and their respective values and/or what input files* were provided for the selected tool, (4) *what output files* were produced and (5) *where* the files were located, (6) *who* ran the simulation job, and (7) *how long* it took to complete the simulation job, and more. The collected provenance record is structured in a JavaScript Object Notation (JSON) format for flexible schema. The JSON document is delivered into the repository.

The repository is implemented as a NoSQL data system. In our work, we adopt MongoDB [19] for the two main reasons. One is that it allows an arbitrary schema to

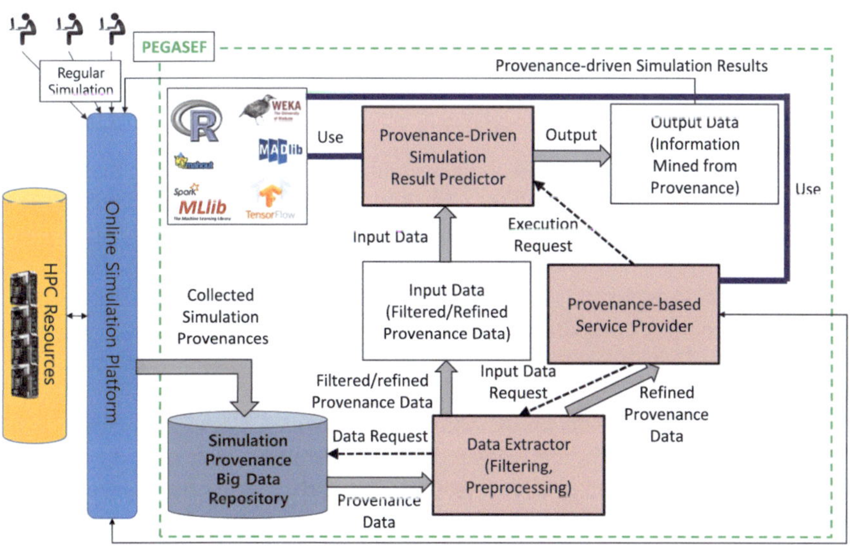

Fig. 2 The proposed provenance-driven big data framework

be specified. There is no need to define a document structure ahead. Since there could be more information (e.g., the number of nodes used for a job, central processing unit (CPU) utilization, etc.) to be collected later, using an unstructured data model makes more sense. Second, MongoDB provides an automatic *sharding*, which is a technique to distribute a huge massive of data across multiple machines. As provenance records can be drastically piled up due to a number of simulations by numerous, simultaneous users, the sharding feature is very useful with little administrative overhead except simply adding a new node whenever needed.

Provenance-based Service Provider: This component is designed to respond to users' simulation requests. If a user submits his/her simulation job to the platform, the Provider component is invoked. The main task of the Provider is to check if a given job matches any existing provenance record. If so, then the new job does not have to be actually executed. Instead, the output files associated with the matching simulation provenance can be simply returned to the user. In that way, we can "avoid" running simulations from scratch. Otherwise, the user is further asked to estimate its simulation result with no actual execution. If this is the case, the Provide sends a prediction request to the Predictor component to perform result prediction.

In addition, the Provider component offers a service of estimating simulation completion time. Namely, it attempts to predict how long it would take to finish the simulation with specified input parameter values/files. For doing the estimation, the Provider utilizes provenance data associated with the tool on which the simulation is conducted. More specifically, it trains via machine learning libraries a model based on the accumulated provenance records. To achieve better accuracy, the Provider also keeps the up-to-date model by retraining periodically based on new provenance data.

Furthermore, the Provider presents another service to cluster existing input parameter sets for given output data that a user uploads. This clustering is particularly useful when a user does not know what input parameter values should be initially set for the output data that she already obtains. This service is provided through provenance data mining.

Data Extractor: This component receives an input data request from the Provider component and then extracts from the repository provenance data associated with the request. It then refines the provenance data by removing any unnecessary parameter or drawing derived values based on raw values. In turn, it supplies the refined data to the Predictor component. This component plays a role in bridging between the Provider or Predictor component and the repository.

Provenance-Driven Simulation Result Predictor: This component responds to a prediction request from the Provider. That is, it predicts simulation results by running machine learning algorithms on the refined data provided by the Extractor component. The predicted results are delivered back to the user. After that, the user can decide whether to take the predicted results or actually execute the desired simulation on the HPC resources.

These four major components of PEGASEF are tightly coupled and work in conjunction with the online simulation platform.

In sum, we expect that PEGASEF can substantially improve the efficiency of simulations on limited and provide distinguished simulation service for users by estimating completion time and clustering input parameter sets based on simulation provenance data.

4 Conclusion

Recently, we witness that thanks to the advanced IT infrastructure, many CSE researchers and students are utilizing HPC simulations for their research and education. In spite of its popularity, an online simulation platform does not seem to make efficient simulation with better service, without filtering requests that may be duplicate or very similar from many users.

To address this concern, we propose PESASEF, which makes aggressive use of existing simulation records, or provenance data, for duplicate elimination on the same simulation requests and advanced simulation mining service such as completion time estimation and input parameter set clustering. The online simulation platform collaborating with the proposed framework is expected to offer enhanced simulation service and satisfy users' increasing demands for prompt simulation and smart simulation assistance, while saving limited HPC resources.

Acknowledgement This research was supported by Kyungpook National University Research Fund, 2017.

References

1. Suh, Y.-K, Ryu, Hoon, Kim, Hanki, and Cho, Kum Won: EDISON: A Web-Based HPC Simulation Execution Framework for Large-Scale Scientific Computing Software. In: IEEE/ACM International Symposium on Cluster, Cloud, and Grid Computing, pp. 608–612. IEEE Press, New York (2016)
2. The EDISON platform, http://www.edison.rc.kr
3. Simmhan, Yogesh L., Plale, Beth, and Gannon, Dennis: A Survey of Data Provenance in e-Science. In: SIGMOD Record, Vol. 34, No. 3, pp. 31–36. ACM (2005)
4. Herschel, Melanie, Diestelkämper, Ralf, and Lahmar, Houssem Ben: A survey on provenance: What for? What form? What from? In: The VLDB Journal, Vol. 26, Issue 6, pp. 881–906. Springer, Heidelberg (2017)
5. The HUBzero platform, http://hubzero.org
6. nanoHUB, http://nanohub.org
7. DataHUB, https://datacenterhub.org
8. SimulationHub, https://simulationhub.com
9. WebMO, https://www.webmo.net
10. Stevens, Robert D., Robinson, Alan J., Goble, and Carole A.: myGrid: Personalised Bioinformatics on the Information Grid. In: Bioinformatics, Volume 19, Issue suppl_1, 3, pp. 302–304, July 2003
11. Taverna, http://taverna.sourceforge.net
12. Kepler, https://kepler-project.org/

13. Ikeda, R., Park, H., and Widom, J.: Provenance for Generalized Map and Reduce Workflows. In: 5th biennial Conference on Innovative Data Systems Research, pp. 273–283. (2011)
14. Akoush, S., Sohan, R., and Hopper, A.: HadoopProv: Towards Provenance as a First Class Citizen in MapReduce. In: USENIX Conference on Theory and Practice of Provenance, pp. 11:1–11:4. USENIX Association (2013)
15. Amsterdamer, Y., Davidson, S. B., Deutch, D., Milo, T., Stoyanovich, J., and Tannen, V.: Putting Lipstick on Pig: Enabling Database-style Workflow Provenance. In: The VLDB Endowment, Vol. 5, No. 4, pp. 346–357. VLDB Endowment (2011)
16. Apache Pig, https://pig.apache.org/
17. Hammad, R. and Wu, C.: Provenance as a Service: A Data-centric Approach for Real-time Monitoring. In: IEEE International Congress on Big Data, pp. 258—265. IEEE (2014)
18. e-Science Central, https://www.esciencecentral.org/
19. MongoDB, http://www.mongodb.com

Preference-Aware Music Recommendation Using Song Lyrics

Sein Jang, Battulga Lkhagvadorj and Aziz Nasridinov

Abstract Music is closely related to our lives. People enjoy music every day and share or express their emotions through music. Therefore, finding and recommending a favorite song among many kinds of songs is important and difficult at the same time. As the online music platform became popular and many people began to use music recommendations. Many approaches have been proposed to understand the user's music preferences and various factors have been used. Also, various recommendation methodologies have been used, such as content-based (CB) or collaborative filtering (CF). In this work, we propose a recommender system using user's preference that was defined based on the feature extracted from the context information of the song's lyrics. We extract and use the latent-topic in the context of the song lyrics. Our results show that it is efficient to recommend songs based on the preferred singers by analyzing user preferences.

Keywords Recommender systems · Music recommendation · Artist recommendation · Context awareness

1 Introduction

Due to the technological development and the scientific breakthroughs, the recommendation systems are becoming more intelligent and user-centric. Such user-centric systems can be applied to the song recommendation systems. Generally, songs that people listen to can define their personality and their tendencies [1]. Thus, we can assume that people, who have similar musical taste, can be similar to each other. In order to analyze one's musical taste, we should consider many song properties like beat, rhythm, lyric poetry to artist and genre, etc. However, the ambiguity of these properties leads to slow and unrealistic recommendation. In our research, we use lyrics as our main property to recommend songs. Lyrics shows the overall char-

S. Jang · B. Lkhagvadorj · A. Nasridinov (✉)
Department of Computer Science, Chungbuk National University, Cheongju, South Korea
e-mail: aziz@chungbuk.ac.kr

W. Lee and C. K. Leung (eds.), *Big Data Applications and Services 2017*,
Advances in Intelligent Systems and Computing 770,
https://doi.org/10.1007/978-981-13-0695-2_18

acteristics of the song (i.e., atmosphere, content, tendency, artist, etc.). Thus, if we analyze the features of song lyrics, which are preferred by the user in the past, we can recommend songs that are useful to the user.

In this paper, we present a recommendation system based on user preference features that are extracted from the lyrics of the songs. Here, songs are located inside user's playlist. The recommendation system consists of two phases. In the first phase, we analyze the user's preferred song's lyrics to create a user profile. Afterwards, we recommend songs to the user, based on their profile. The main feature of the proposed method is that it can recommend not only songs but also artists to the users based on their profile. We performed experiments that demonstrate the superiority of the proposed recommendation method under various conditions.

The rest of this paper is organized as follows. In Sect. 2, we discuss the related works. Afterwards, in Sect. 3, we demonstrate our motivation. In Sect. 4, we would explain our proposed recommendation method. In Sect. 5, we show our experimental results. Finally, in Sect. 6 we conclude and discuss the future work.

2 Related Works

In this section, we discuss the related works. This section is divided into two subsections. Section 2.1 discusses latent Dirichlet allocation (LDA) [2], the key method used in this study. Afterwards, in Sect. 2.2, we review various song recommendation systems.

2.1 Latent Dirichlet Allocation (LDA)

Latent Dirichlet allocation is one of the topic modeling techniques for extracting topics from corpus. LDA is a probabilistic model of what topics exist in each document for a given document. It estimates the distribution of words per topic in a document, and the distribution of topics per document. To explain the distribution of words per topic, there is a specific topic A, the probability that the word "Lyrics" will appear in this topic A is 0.05, the probability of "bit" to appear is 0.03, and the probability of "rhythm" to appear is 0.02. From these, we can deduce that topic A is a topic related to "songs". More specifically, when the document D is present, the main topic of the document D is likely to be topic A if the proportion of the words belonging to the topic A among the words appearing in the document D is greatest.

In this study, we will use LDA to find hidden topics in song lyrics and use the topics as a feature of the song to create a profile of the song.

2.2 *Song Recommendation Systems*

In past decade, most of the digital contents were transformed and stored in the online platforms. Intelligent music/song recommendation systems became a trending research field. One of the examples of such systems is context-aware music recommendation systems. In [3], authors proposed to use the data of the songs which are tagged in the social tagging website. They argue that it can be reliable song recommendation system. Their approach was to map the tagged song data and discover frequent patterns. Given the user's sequence of contexts, their system can predict the next topic of the playlist. Unlike others, authors of [4] proposed an assumption that the current song recommendation systems were not satisfying the short-term need of the users. In addition, they have assumed that the user's short-term need was mostly influenced by their emotional states, activities, or external environment, based on the studies of psychology and sociology researches. Therefore, the authors conducted a research on music systems based on the activities. The proposed method classifies the daily activities into six different classes, which are Running, Working, Sleeping, Walking, Shopping, and Studying. Although their research shows promising results, it cannot fully achieve the actual goal of the music recommendation system, which is to present the song that can fulfill the musical taste of the user. Based on results of [3], our proposed method analyzes the lyrical poetry to obtain the topics. Moreover, our proposed method is more suitable for fulfilling the users need through a recommendation system unlike [4]. In addition, authors of [5] proposed music recommendation system to solve the gray sheep user problem. Gray sheep users are the group of users with unusual preferences. In some cases, those users are called outliers. Usually, such users receive poor recommendations from the system as they do not have the same interest as the other users. Thus, their system is based on the use of implicit information to obtain user preferences such as the number of plays, which artist they are fan of. Their experiments and studies show promising results, however, the proposed method only makes recommendations based on songs in the playlist. However, in our proposed approach, we can mitigate the problem of gray sheep users of the system, because our approach uses the main property of the song, the lyrics, and thus, our approach has better accuracy.

3 Motivation

In this section, we demonstrate the motivation of our work. People create their own personal playlist based on their musical taste. In addition, they prefer to create their playlist by songs of certain artists, due to their own unique style of rhythm and lyrics poetry. Therefore, if we classify people by their playlist, people with same musical taste would be classified together. Using such information, we could recommend songs, which they might prefer but which are not in their playlist.

In our work, among many components (e.g., tune, lyrics, beat) of songs, we chose lyrics as our main feature. Lyrics contribute a lot to the overall atmosphere of the song. Artists use their lyrics to convey their thoughts and values to the public. All these points taken into consideration, by analyzing a user's playlist, we could obtain information, which contains the musical taste and the artists who are preferred by the user. Therefore, in this paper, we perform topic-modeling on the lyrics. We use latent Dirichlet allocation (LDA) topic modeling approach to represent songs by their lyrics. The LDA estimates the distribution of words per topic, and the distribution of topics per document which is lyrics in our work. We generated LDA model with 25 topics. The data used to generate the LDA model has 47,104 lyrics from LyricsFreak. Using LDA model, we obtained relevant words to each topic. The results are shown in Table 1.

In a similar way, we can obtain latent features of artists. All the songs are classified into 25 topics in the LDA model. Afterwards, we compare the songs of the artist with the 25 topics, which we described earlier. For example, in case of the Kanye West, 58% of his songs were more relevant to the topic 6. Eminem, another hip-hop artist like Kanye West, has also his songs in topic 6 with 74%. Table 2 lists the top 3 topics for six artists from three different genres.

As we can see from Table 2, we derive the latent features of each artist through the LDA model. Furthermore, we can make an assumption that artist from the same genre falls into similar topics.

Figures 1 and 2 show visualization of topics estimated using latent Dirichlet allocation by LDAvis [6]. It shows global view of the 25 topics that modeled by 47,104

Table 1 Most relevant words to each topic

Topic 1	Topic 2	Topic 3	Topic 4	…	Topic 22	Topic 23	Topic 24	Topic 25
know	love	life	time		rock	music	la	shake
want	heart	world	back		roll	party	da	de
make	never	u	day		beat	rhythm	beautiful	le
got	ill	one	home		blood	send	an	bang
thing	know	na	long		n	saturday	ba	pum
way	one	hand	gone		soul	play	monday	pa

Table 2 Topic and artists

Artists	Genre	Top-1 topic	Top-2 topic	Top-3 topic
Kanye West	Hip-Hop	Topic 6 (58%)	Topic 8 (18%)	Topic 21 (3%)
Eminem	Hip-Hop	Topic 6 (74%)	Topic 8 (14%)	Topic 18 (2%)
Adele	Pop	Topic 5 (37%)	Topic 8 (31%)	Topic 9 (9%)
Mariah Carey	Pop	Topic 5 (34%)	Topic 8 (20%)	Topic 10 (10%)
Nirvana	Rock	Topic 8 (30%)	Topic 18 (10%)	Topic 22 (4%)
Linkin Park	Rock	Topic 8 (36%)	Topic 9 (13%)	Topic 21 (6%)

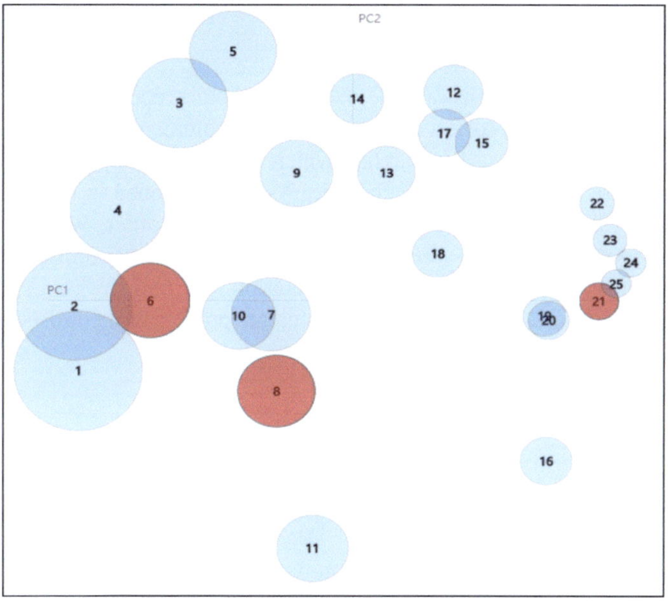

Fig. 1 Top-3 most relevant topic of "Kanye West" was marked in red. Topic 6 (58%), Topic 8 (18%), and Topic 21 (3%)

lyrics data. It is the result of projecting a high-dimensional result (i.e., 25 dimensions in our case since we modeled it with 25 topics) of the topic modeling on the two-dimensional plane. The areas of the circles are proportional to the relative prevalence of the topics in the words. The distances of topics are computed by Jensen-Shannon divergence. With this visualization tool, we can easily see the relationship between topics.

4 Proposed Method

In this section, we demonstrate our proposed method. Our recommendation method consists of two phases. In the first phase, we profile the artists and songs based on their lyrics. Afterwards, we make our recommendation based on the profiles which we have created in the previous step by using the user's playlist. In Sect. 4.1, we explain the overview of the proposed method. Sections 4.2 and 4.3 demonstrate the functionalities of the artist and the song recommendation separately.

4.1 Overview

In this subsection, we demonstrate the overview of our proposed method. Figure 3 illustrates the abstracted process of our proposed recommendation. First, the input data of our method is the user's playlist, which later would be transformed by the LDA to compare the values with each topic. As we mentioned in the beginning of Sect. 4, this is the profiling phase. Next, by using the profiles, we can make the recommendations to the users who have the same interest. Table 3 lists the frequent notations, which are used in our further explanation.

In our study, we use user's playlist data to create profiles of the artists and songs. Profile of artists A_i is denoted as follows:

$$A_i = \{A_{i,n} = (S_{i,n}/S_{i,\text{total}}), n = 1, \ldots, N\} \tag{1}$$

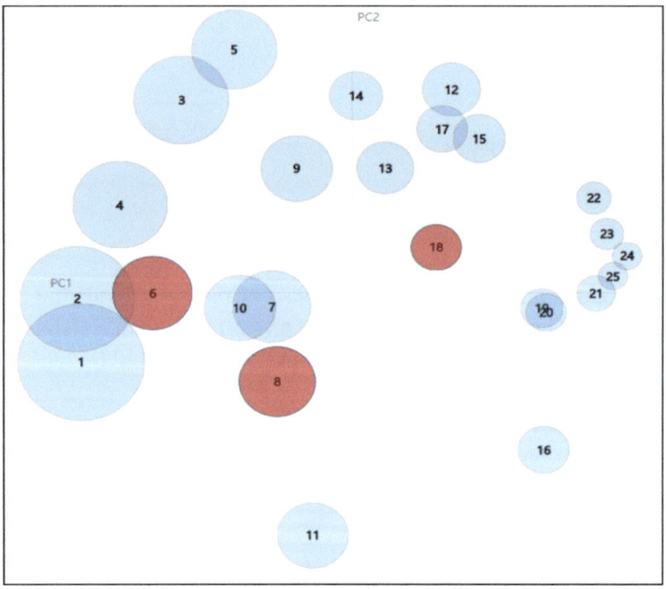

Fig. 2 Top-3 most relevant topic of "Eminem" was marked in red. Topic 6 (74%), Topic 8 (14%), and Topic 18 (2%)

Table 3 Notations

Notation	Definition
N	Total number of topics
A_i	ith artist ($1 \leq i \leq N$)
S_j	jth song ($1 \leq j \leq N$)

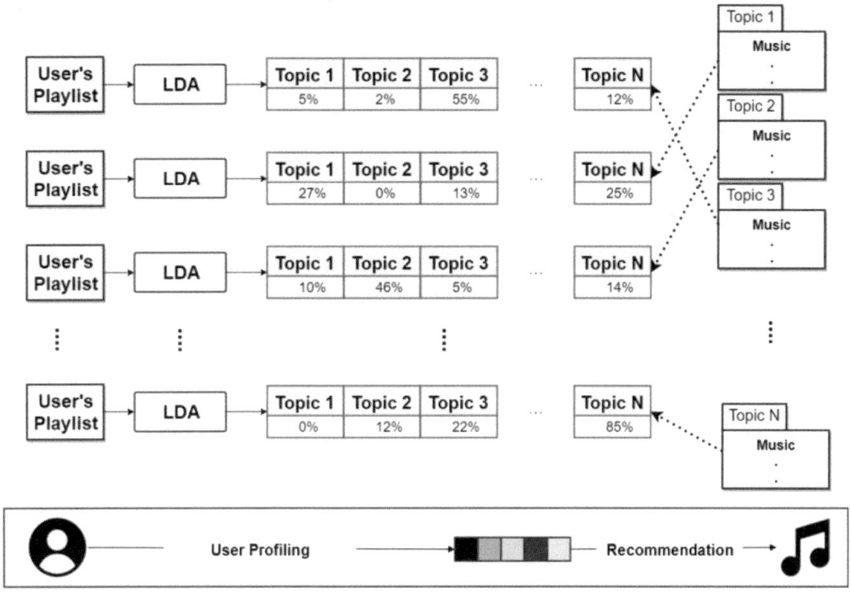

Fig. 3 An overview of proposed recommendation method

where $A_{i,n}$ represents instance of the A_i correlation with topics. $S_{i,n}$ is number of songs in topic n. $S_{i,\text{total}}$ is total number of songs for A_i. The profile of the A_i is obtained by dividing the number of songs in each topic by the total number of songs of A_i. N represents total number of topics.

In case of songs, the profile of song S_j is a set of instances denoted as:

$$S_j = \{S_{j,n}, n = 1, \ldots, N\} \tag{2}$$

where $S_{j,n}$ represents instance of the correlation between S_j and topics. N represents total number of topic.

Example 1 In this example, we calculate the profile of the artist A_1. A_1 has 10 songs in total and 6 of its songs belong to topic-1, 3 of its songs belong to topic-2 and its last song belongs to topic-3. The result of profile calculation is shown in Table 4.

As we can see from the result, the profile of A_1 could be represented as a vector with length of N with corresponding values.

Given U users and I artists, J songs, we use artist profile $\text{Artist}_i (i = 1 \ldots I)$, and song profile $\text{Song}_j (j = 1 \ldots J)$, and user playlist $\text{Playlist}_u =$

Table 4 A_1's profile

Topic-1	Topic-2	Topic-3	Topic-4	…	Topic-N
60.0	30.0	10.0	0.0		0.0

$\{(\text{Artist}_1|\text{Song}_1), (\text{Artist}_1|\text{Song}_2), \ldots, (\text{Artist}_I|\text{Song}_2), \ldots, (\text{Artist}_I|\text{Song}_J)\}$. The goal of the recommendation is predicting a song that the user will like among the songs not in the user's playlist. In summary, the contributions of this work are as follows:

1. We perform topic modeling using lyrics. Using lyrics, we obtain the latent features of a song or artist and analyze their association.
2. Using latent features from lyrics, we recommend artist to user, based on user's playlist. We can also determine the user's preference for artist.
3. By considering user's preference for artist, we recommend songs to the user. The experiments show that results of the song recommendation using user's preference for artist show better performance than using only song's features.

4.2 *Artist Recommendation*

We obtain artists' profiles based on topic modeling. Using profiles and user's playlist, we recommend artists to the users, who are unknown to the user, yet user would appreciate the songs of that particular artist. To generate artist's profile, we consider each song of the artist to determine which topic is most relevant to the artist. For example, Kanye West has a total of 106 songs and among them, 63 songs are more relevant with topic-6. Accordingly, we can calculate the relevance by $(63/106) * 100 = 59.433962$. By performing these calculations for all topics, we can generate the artist's profile.

Furthermore, we use two methods to recommend the artist. At first, we recommend an artist, who is more relevant with artists that are already in user's playlist. We obtain correlation between artists by comparing their profile. Thus, Pearson correlation coefficient was used. Second, we recommend Top-N artists that are more relevant to topic, which are already in user's playlist.

4.3 *Song Recommendation*

For generating song profile, we use the LDA model with topic modeling by using lyrics. We obtain the distribution of the topics that each song has and use it to generate song profile. Since the LDA model is generated with 25 topics, the song profile becomes a one-dimensional vector with length of 25 with corresponding values. Each value is the ratio of each topic to the song's lyrics.

For song recommendation, we used three recommendation methods. First, we consider the correlations between songs. We obtain Pearson correlation coefficient between all songs and recommend songs of top-N. Second, we consider both the correlations between artists and songs. We recommend top-N songs from values obtained by multiplying correlations between artists and correlations between songs. Finally, we consider both artists and songs, but we give priority to artists, who can make an impact on the user's preference. We list up the artists that are correlated with the artists in the playlist of the user. Afterwards, we choose song, which should be recommended to the user based on the correlation value.

5 Experiment

In this section, we present experimental results. In these experiments, we used data that consists of artists, songs, and lyrics. Such datasets are difficult to generate. Thus, we used playlist datasets [7] that are collected by Cornell University. The playlist dataset is crawled from Yes.com and Last.fm. In addition, lyrics dataset was downloaded from LyricsFreak [8]. After performing the preprocessing on both datasets, we obtain a total of 460 artists with 47,104 songs (with lyrics) and 11,138 playlist data.

5.1 Experiment Result

For the experiments, we randomly chose one playlist, which has 20 songs. We assume that we know one of the songs from the playlist. The randomly chosen song was "Kanye West—Runaway". At first, for artist recommendation, we generate artist profile based on our dataset. We perform topic classifications on songs and obtained artist profile. Sample of topic classifications for Kanye West's songs is shown in Table 5, and Table 6 shows Kanye West's profile.

In order to use the two recommendation methods that we mentioned earlier, we sought correlation between Kanye West and other artists, and the correlation between

Table 5 Topic classifications for songs

Artist	Song	Most relevant topic
Kanye West	I Just Wanna Love U	Topic 1
	I Love Kanye	Topic 12
	I'll Fly Away	Topic 21
	Late	Topic 6
	Runaway	Topic 6
	Higher	Topic 6

Table 6 Kanye West's profile

1	2	3	4	5	6	...	25
0.943396	0.0	1.886792	1.886792	2.830189	59.43396		0.943396

topic-6, which is the topic that is more relevant with “runaway”. The results are shown in Tables 7 and 8.

We could see from Table 7 that there is a high correlation between artists who have similar tendencies.

The results of the recommendations using the two methods described above which consider correlation between artists and consider topic relevance are shown in Figs. 4 and 5. In the graph, the precision is the ratio of the number of artists found in the user’s playlist that the user prefers, among the recommended artists. The recall is the ratio of the users’ preference to artists, that is, the number of artists actually found among the artists in the user’s playlist. For example, if a user’s playlist has 10 artists, 100 artists are recommended, and 5 artists are artists in the user’s playlist, the precision is $5/100 = 0.05$ and the recall is $5/10 = 0.5$.

As can be seen from the two graphs, recommendation using correlation between artists shows better performance. We recommended the artist to the user, so that we know what kind of artist user likes and what kind of genre user likes, and furthermore, we can infer the user’s preference for song. Now, using the user’s preferences, we recommend song as well as recommending artists.

For song recommendation, we generate a profile for a total of 47,104 songs using the method described above. Profile of Kanye West’s Runaway is shown in Table 9. Here, we used three recommended methods. The results of three methods are shown in Fig. 6. The method that considers user’s preference shows better performance than other methods.

Table 7 Correlation between Kanye West and other artists

Artist	Pearson’s correlation coefficient
Eminem	=0.98
Drake	=0.94
Snoop Dogg	=0.96
Michael Jackson	=0.12
Queen	=0.10
Bon Jovi	=0.10

Table 8 Top-15 artists with the most songs related to topic 6

	Topic-1	Topic-2	…	Topic-6	…	Artist
1	0	0		100		YG
2	0	0		100		Young Buck
…						
13	0	0		=91.54		Snoop Dogg
14	0	0		=88.67		Wu-Tang Clan
15	0	=1.60		=86.40		Lil Wayne

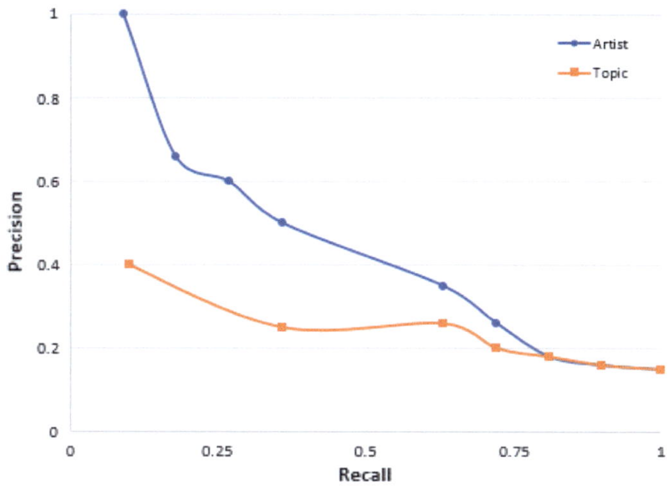

Fig. 4 Precision-recall curves for different artist recommendation algorithms. Recommendations using correlations between artists outperformed the recommendations using correlations between topics by 90% in average precision

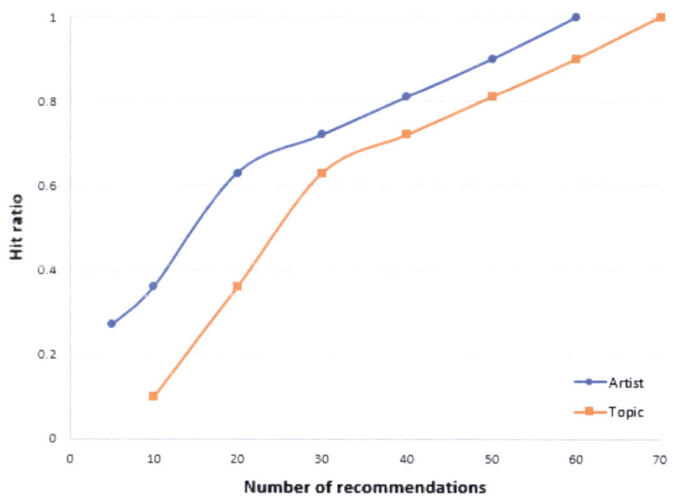

Fig. 5 Hit ratio for different number of recommendations of artists

Result of correlation between "Runaway" and other songs that calculated by Pearson's correlation coefficient shown in Table 10.

Each graph represents three recommended methods. "Song" considers only the correlation of song. "Artist * Song" considers both song and artist correlations. "Artist" considers only the correlation of artists.

Table 9 Kanye West—Runaway profile

1	2	3	4	5	6	…	25
0.046665	0.0	0.063207	0.044414	0.031877	0.212388		0.04233

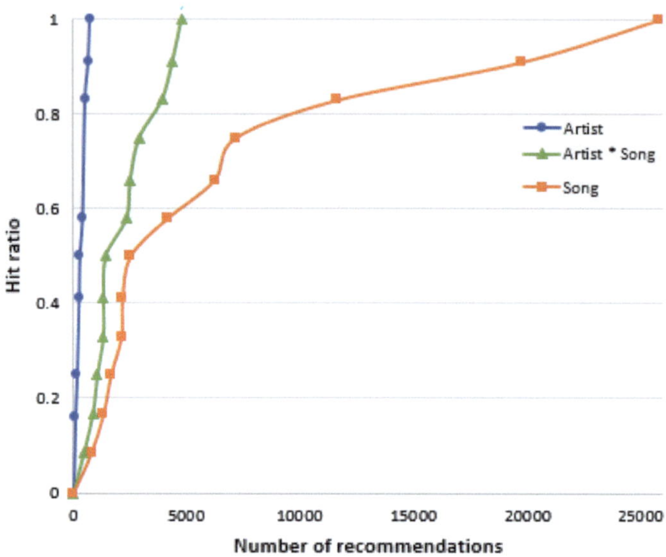

Fig. 6 Hit ratio for different number of recommendations

Table 10 Correlation between Kanye West—Runaway and other songs

Artist	Song	Pearson's correlation coefficient
Wiz Khalifa	Remember You	=0.92
Drake	We'll Be Fine	=0.91
Z-Ro	Hey Lil Mama	=0.91
Eminem	Careful What You Wish For	=0.90
Ying Yang Twins	My Brother's Keeper	=0.90
Jennifer Lopez	Acting Like That	=0.90

6 Conclusions and Future Work

We have presented personal preference-aware song recommendation, a song recommendation model that considers user's preference which is artist in our study. We proposed a topic modeling based on lyrics for song recommendation. Using that model, we generate artist profile and song profile. To evaluate our recommendation, we used a total of 460 artists of 47,104 songs (with lyrics) and 11,138 playlist data.

Our recommendation results show that our song recommendations system has better performance.

In the present work, we modeled topics using only lyrics, but there are many other features in songs other than lyrics such as rhythm and beat. So, in future work, we plan to apply more features to topic modeling. In addition, the preference of the user is not limited to the singer, therefore the genre of the song and the tendency of the person will also be considered.

Acknowledgement This work was supported by Institute for Information & communications Technology Promotion (IITP) grant funded by the Korea government (MSIP) (No.2016-0-00406, SIAT CCTV Cloud Platform).

References

1. Schwartz, K. D., Fouts, G. T.: Music preferences, personality style, and developmental issues of adolescents. In: Journal of youth and adolescence, 32(3), 205–213 (2003)
2. Blei, David M., Andrew Y. Ng, and Michael I. Jordan.: Latent Dirichlet allocation. In: the Journal of machine Learning research. 3: 993–1022 (2003)
3. Hariri, N., Mobasher, B., Burke, R.: Context-aware music recommendation based on latent topic sequential patterns. In: proceedings of the sixth ACM conference on Recommender system, pp. 13–138. ACM (2012)
4. Wang, X., Rosenblum, D., Wang, Y.: Context-aware mobile music recommendation for daily activities. In: Proceedings of the 20th ACM international conference on Multimedia, pp. 99–108. ACM (2012)
5. Sánchez-Moreno, D., González, A. B. G., Vicente, M. D. M., Batista, V. F. L., & García, M. N. M.: A collaborative filtering method for music recommendation using playing coefficients for artists and users. In: Expert Systems with Applications, 66, 234–244 (2016)
6. Sievert, C., Shirley, K.: LDAvis: A method for visualizing and interpreting topics. In: Proceedings of the workshop on interactive language learning, visualization, and interfaces, pp. 63–70 (2014)
7. Playlist Dataset-Cornell Computer Science—Cornell University, http://www.cs.cornell.edu/~shuochen/lme/multilme_page.html
8. http://www.lyricsfreak.com/

Real-Time Smart Safe-Return-Home Service Based on Big Data Analytics

Gae-A Ryu, Jae-Won Lee, Ji-Sung Jeong, Mihye Kim and Kwan-Hee Yoo

Abstract In modern society, various forms of crime are constantly occurring. Accordingly, several safe-return systems for the socially vulnerable are being developed. However, those systems are mainly focused on responding to dangerous situations that have already occurred, and they do not predict the possibility of crime reflected by information about the user's surroundings in real time. This paper proposes a new safe-return-home service that allows users to be notified of, and therefore handle, the possible dangerous situations surrounding them in real time. This is accomplished by collecting and analyzing various types of big data about the user's surroundings in real time. Collected and analyzed data include the locations of users, the locations of CCTV (Closed-Circuit Television) cameras, crime/disaster/accident-related real-time news data, the locations of shelters, real-time CCTV video data, and social network service data. Through the analysis of these data, the prediction of potential surrounding dangers is visualized on user devices, and ideas for counteracting those dangers are suggested to users in real time.

Keywords Safe-return service · Safe-return-home service · Big data analytics

---

G.-A. Ryu · J.-W. Lee · J.-S. Jeong · K.-H. Yoo (✉)
Chungbuk National University, Cheongju, Republic of Korea
e-mail: khyoo@cbnu.ac.kr

G.-A. Ryu
e-mail: reflexive@gmail.com

J.-W. Lee
e-mail: lnodie86@gmail.com

J.-S. Jeong
e-mail: farland83@cbnu.ac.kr

M. Kim
Catholic University of Daegu, Daegu, Republic of Korea
e-mail: mihyekim@cu.ac.kr

W. Lee and C. K. Leung (eds.), *Big Data Applications and Services 2017*,
Advances in Intelligent Systems and Computing 770,
https://doi.org/10.1007/978-981-13-0695-2_19

1 Introduction

In today's era of smart technology, thanks to the development of information and communication technology (ICT) and the popularization of the smartphone, data are constantly being transmitted between devices and users and between devices and other devices. The amount of data produced is geometrically increasing, and the environment has now become one where most systems have to use big data. Based on such trends, various services for user convenience are being developed on smart devices based on big data [1]. A safe-return service is one such application system [2].

In modern society, various forms of crime are constantly occurring, and in particular, crime aimed at the socially vulnerable, such as women returning home late at night, the elderly, the weak, and youths, is frequent, which has contributed to social anxiety. Accordingly, many safe-return-home services [3–8] for the socially vulnerable are being developed and provided. These services, however, are mostly focused on responding to dangerous situations that have already occurred and do not include services that inform users about potentially dangerous situations to cope with in advance by analyzing situations in the user's surroundings.

In this paper, we propose a safe-return service that notifies users about possibly dangerous situations through real-time collection and analysis of various sets of big data regarding the user's surroundings, including the locations of users, the locations of CCTV cameras, crime/disaster/accident-related real-time news data, the locations of shelters, real-time CCTV video data, and social network service (SNS) data. By collecting and analyzing such data, the system visualizes and presents the surrounding dangerous situations to users in real time. The content of this paper is mostly based on work presented in a previous paper [9, 10].

The remainder of this paper is organized as follows. In Sect. 2, we review the existing literature about safe-return services. In Sect. 3, we describe the overall configuration and major features of the proposed safe-return service. In Sect. 4, we present the implementation results of the service on a smartphone, and Sect. 5 concludes the paper with some directions for future research.

2 Related Works

The most prominent service for a safe-return is the SOS National Safety Service provided by the Ministry of Security and Public Administration (MOSPA) [3]. This system responds to crimes by searching for the profile and location of a user when the user presses the shortcut key (112) in a dangerous situation. The retrieved profile and location are used to instruct the nearest patrol car to proceed to the user's location to prevent the crime there. However, it may be difficult for a user to press the shortcut key in a dangerous situation. Hence, only a small number of users are currently utilizing this service.

Another system provided by MOSPA is Smart Safe-Return [4]. This is a location-based app in which users can register guardians with their phone numbers or emergency contacts, and they can set a default contact (first guardian). In addition, they can register destinations and set alarm types and intervals. When a user runs this app on his/her smartphone, the system delivers the current location of the user to the default guardian in real time until the user reaches his/her destination. The alarm can be set in 1–60-min intervals, and the alarm types include text messages as well as Facebook, Twitter, and KaKaoTalk messages. Safe and dangerous regions can also be set to notify the guardian when the user reaches a certain region. There is also a button that can directly connect to the police station or a fire station during service progress, and there are other additional features, such as the indication of safe facilities in nearby areas. However, this service cannot predict user surroundings in real time, which would make it possible to predict the possibility of crime.

The Safe-Return Project app [5] is another safe-return service that executes a 'real-time safe-return sensor' feature on a smartphone during a commute to work or school, which rapidly delivers a text message and makes a phone call to a preset number in case of an emergency. After running the real-time sensor, when the system recognizes three rapid touches of the smartphone screen, three or more shakes of the smartphone, or a loud scream from the user, the system delivers a text message containing the current location and automatically dials the preset number for an emergency situation. If a user has been in the same location for longer than a certain period of time, the system also recognizes this as an emergency situation. During the service, the system automatically stores the route of the user's movements as well as voice recordings, and it provides an icon on its main screen to enable direct connection to 112. However, the service cannot forecast dangerous situations and regions that might exist in the route of the user.

Elegant Homecoming [6], provided by an individual developer as an Android App for Google Play, is another emergency rescue app that has an elegant design for women's preferences. Users can make a fake telephone call in a dangerous situation using recorded conversations. During the fake call, the users can send an emergency signal to a registered number by touching a button on the system screen. The routes of users are stored in real time to check whether they are returning home safely or not; if the users are out of the routes, a signal is sent to a registered number. The default preset number is 112. In emergency situations, users also can directly deliver an alarm to the preset number. This app also provides the information of the safe-return accompaniment service operated by the Seoul metropolitan government. However, it does not support a service that indicates potential risks and dangers surrounding the user. There are other, similar systems, such as the Taxi Safe U [7] and Help Me (SOS Emergency Alarm) [8] services, but all of them only focus on broadcasting current dangerous situations that users face; none of them can predict the possibility of danger by analyzing the user's surroundings, to allow users to come up with proper countermeasures against the dangers.

3 Safe-Return Service

3.1 System Configuration

Figure 1 shows the overall process of the proposed safe-return service. When a user sets a destination after logging on to the system, the server calculates the distance from the user's current location to the destination, and sets an alarm based on the estimated arrival time at the destination. While the user moves toward the destination, the user's location is periodically sent to the server. Upon the occurrence of an unexpected situation, for instance, when the timer ends before the user arrives, the server sends alarm messages sequentially in three steps. In the first step, a message is sent to the user. If the user is safe, he/she is requested to restart the system. If there is no response from the user in the first step, the system informs the possibility of danger to a registered guardian in the second step. If there is no response from the user after the second-step notification, the system regards this as an emergency situation and sends the user information to a nearby police or fire station [9].

Figure 2 shows the main features of the server for the safe-return service. The server is largely composed of three features: data collection, data analysis, and notification. In data collection, the following data are collected: surrounding information (shelters, safety regions, locations of CCTV cameras), location data (using a global positioning system [GPS], base station [BS], and Wi-Fi/3G), the video data of the CCTV camera, disaster/accident/crime-related real-time news, SNS data, and existing crime data. The collected data are analyzed in terms of the following: user

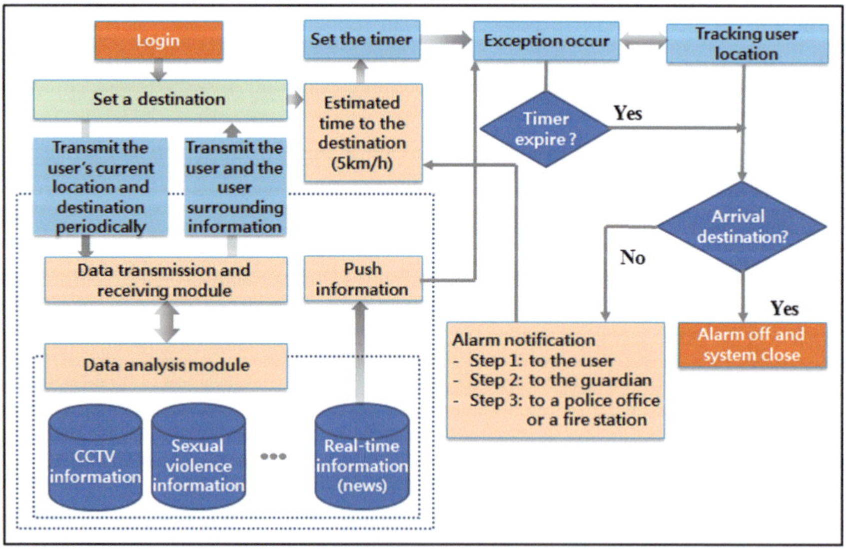

Fig. 1 Process of the proposed safe-return service

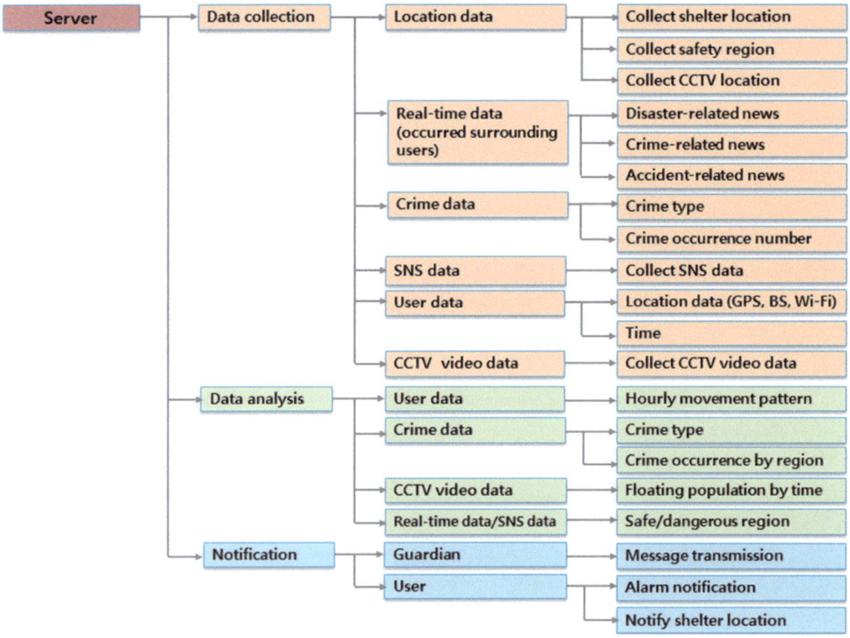

Fig. 2 Main features of the server for the safe-return service

movement patterns at regular time intervals based on the user information, the types of crimes based on the criminal data, and the flowing population by time using the CCTV data. Then safe and dangerous regions are identified based on the online news and SNS data in real time. Users are notified of the locations of shelters and dangerous regions in their surroundings, or a text message is sent to the default guardians appointed by the user.

Figure 3 presents the main features for clients of the safe-return service. The features consist of notification, retrieval, visualization, analysis, and feedback. When an emergency situation (disaster, crime, or accident) occurs, the client system informs users about the situation, and provides information on where they can go, along with the locations of nearby safe regions (e.g., shelter, police station, fire station). Through the client system, users can also directly search for safe regions, as well as for the locations of nearby pharmacies and hospitals.

The client system also visualizes information in the user's surroundings. For example, the disasters, crimes, and accidents that have occurred in that specific area are visualized using graphic tools. The system also analyzes the average speed of the user's walking routes and preferred modes of transport using their movement patterns. Accidents are determined using a motion sensor and the user's situations are apprehended in connection with other relevant apps. For instance, using the near field communication (NFC) tags of taxi apps, the system identifies taxis the user has taken and where the user is headed. If the user stops (i.e., if the GPS location

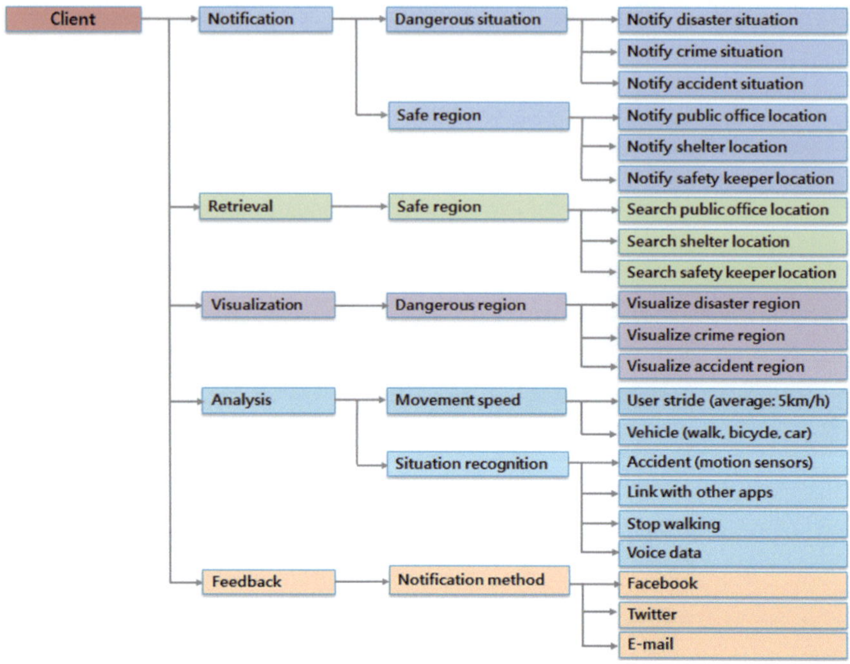

Fig. 3 Main features of clients for the safe-return service

information does not change), the recording function of the user's smartphone is activated to determine whether the user is in a dangerous situation, and the recorded voice data are analyzed to further assess the current situation. The client system also allows users to send a personal SNS or an e-mail to notify their friends of their situation [9].

3.2 *Main Features*

3.2.1 Data Collection

The big data used for the safe-return service are categorized into offline and online datasets as shown in Fig. 4. Table 1 describes the fields of each of those datasets.

User data are collected from the location of the user's smartphone and are managed by dividing them into two categories: login information and location information. The location information is periodically collected whenever the user moves over 100 m. This information is used to analyze the speed of the user walking on foot or using a preferred mode of transport. We used the CCTV data provided by a portal website for the public (www.data.go.kr). We calculated the latitude and longitude

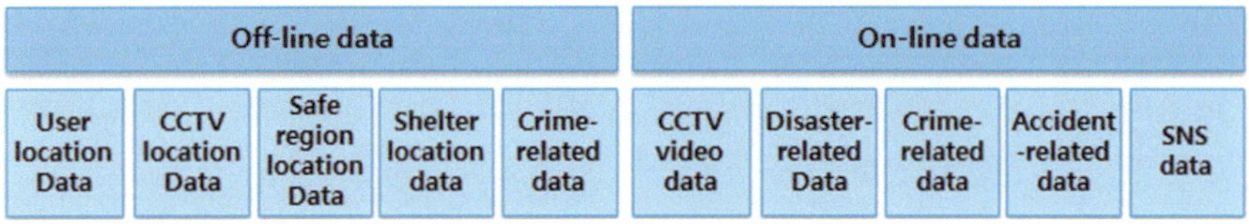

Fig. 4 Offline and online database used in the safe-return service

Table 1 Fields of each dataset used in the safe-return service

	Data type		Fields
Off-line data	User data	Login information	User id, password, name, address, phone number, and device number
		Position information	Index, device number, latitude, longitude, and date
	CCTV data		Department in charge and installation number, objective, and address
	Public office data		Office name, address, type, latitude, longitude
Online data	News data related to accidents, disasters, and crimes		Title, subtitle, text
	CCTV video data and SNS data		Text

of the user location from the installation addresses of the CCTV cameras, using geocoding. The 'objective field' of the CCTV dataset indicates the purpose of each CCTV installation, such as river surveillance, traffic information collection, disaster management, and so forth [9].

We imported the location information for public offices such as police and fire stations, shelters, and safety keepers from the Smart Safe-Return system provided by MOSPA [4]. The disaster/crime/accident-related real-time data are collected from an online rich site summary (RSS) of Chungbuk Ilbo (a local newspaper) using a crawler we developed that collects news data. Figure 5 indicates each process for collecting the locations of public offices from the Smart Safe-Return service and the news data from Chungbuk Ilbo [9].

3.2.2 Data Analysis

Data analysis is conducted for real-time data. Figure 6 presents the overall process of data analysis in the proposed safe-return system. The collected real-time data are stored in a Microsoft SQL (MS SQL) server database and analyzed using Apache Hadoop MapReduce [11–13]. The analyzed data are stored in the Hadoop file system (HDFS) and then transmitted back to the MS SQL server.

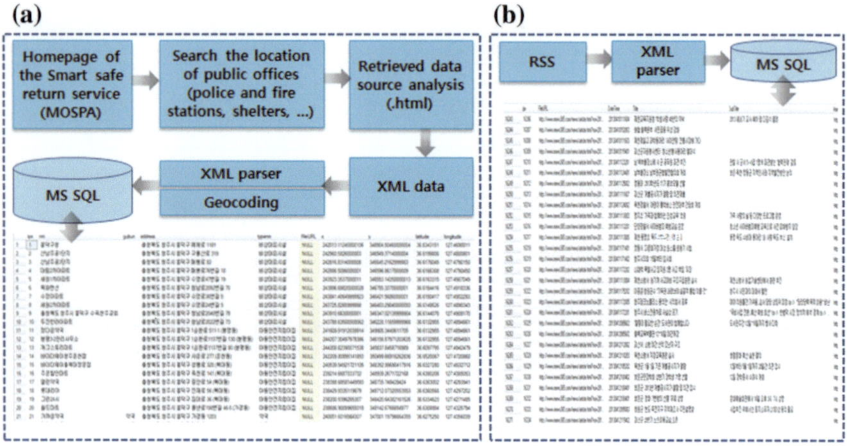

Fig. 5 Process for collecting **a** the location of public offices and **b** news data

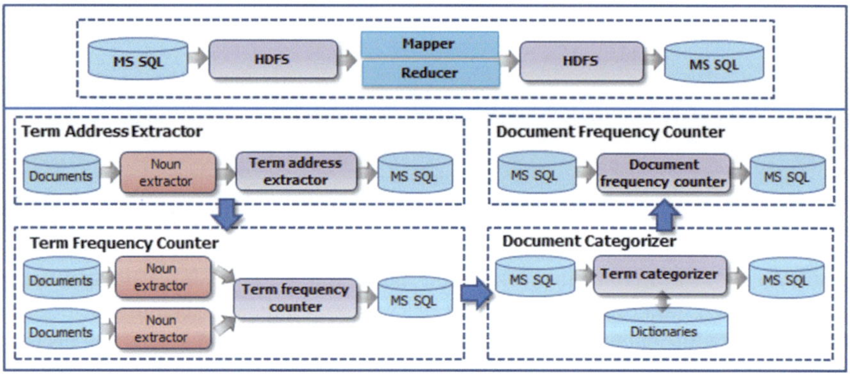

Fig. 6 Overall process of data analysis

The real-time data are mostly text data, and thus, natural language processing is necessary. For this purpose, we used the HanNanum Korean lexical analyzer provided by the Korea Advanced Institute of Science and Technology (KAIST); that is, we analyzed the text data (documents) parsed into morphemes using the HanNanum Korean lexical analyzer [14]. Then, we divided the morphemes into noun units and extracted address data from the noun units by comparing them to the address dictionary [15–17]. We built address, crime, accident, and disaster dictionaries and then calculated the frequency of related terms for each unit of extracted data. Then, we categorized the documents using the term frequency-inverse document frequency (TF-IDF) algorithm [18] and calculated document frequency for all extracted data; for example, if the term "sexual violence" appeared in two documents, then the document frequency for the term will be two. The latitude and longitude at which extracted data were obtained were determined using geocoding.

3.2.3 Data Visualization

The analyzed data are visualized via Google Maps. We used Google's crime and disaster icons to visualize pertinent data, as shown in Fig. 7. Icons of public offices were designed as shown in panel (c) of that figure. The size of each icon is different depending on its term frequency and is changed dynamically. When an icon is clicked, more detailed information appears. Figure 8 shows an example of the location of a user obtained through the GPS and Wi-Fi/3G functions of a user's smartphone; panels (b) and (c) show visualization examples for the public office and CCTV camera data, respectively. We used free icons from Iconfinder (https://www.iconfinder.com/free_icons) for the visualization of the CCTV data.

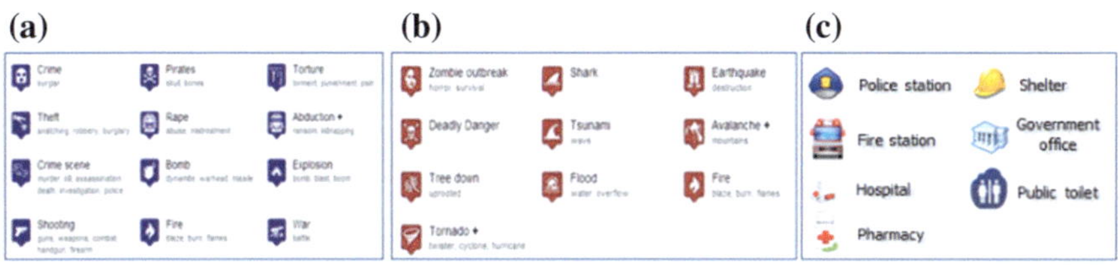

Fig. 7 Icons used for **a** crime, **b** disaster, and **c** public office data

Fig. 8 Visualization examples for data on **a** a user position, **b** public offices, and **c** CCTV cameras

Table 2 Implementation environment of the safe-return service

Server–Client		Specification
MS SQL Server	CPU	Intel(R) Xeon(R) E5345 @ 2.33 GHz
	RAM	8 GB
	OS	Windows Server 2008 R2 Enterprise
Hadoop Server x 4	CPU	Intel(R) Core(TM) i5-3470 CPU @ 3.20 GHz
	RAM	16 GB
	HDD	3 TB
	OS	Linux version 2.6.32-431.e16.x86_64
	SW	Hadoop 1.2.1, Java 1.7.0_45
Client	CPU	2.33 GHz Quad Core
	RAM	2 GB
	OS	Android 4.2.2
	Device	Samsung Galaxy S4

4 Implementation Results

Table 2 indicates the implementation environment of the safe-return service. The collected real-time data are stored on the MS SQL server and analyzed on the Hadoop server. The analyzed data are saved back to the MS SQL server and then visualized on a client device.

Figure 9 shows the composition of the initial screen of the system that consists of five buttons: "favorites," "view settings," a quick button for "safe regions," "SNS," and "settings." Users can add or delete a destination to the system using the "favorites" button, and they can select facilities to be displayed on the map through the "view settings" button. Using the "settings" button, users can also register the phone numbers of guardians and decide whether or not they will automatically report their situation to the police station (112) or the fire station (119) when an emergency occurs. The "SNS" button sets up posting the user's location information on Facebook [9].

Figure 10 shows visualization examples of safe regions: (a) fire stations, (b) police stations, and (c) hospitals. Other safe regions include pharmacies and shelters. Using the "safe regions" button, users can select the type of safe regions to display on the map.

Figure 11 shows visualization examples of dangerous regions: (a) all dangerous situations (crimes, disasters, and accidents), (b) disasters (fire), and (c) accidents. These data are collected from real-time news services.

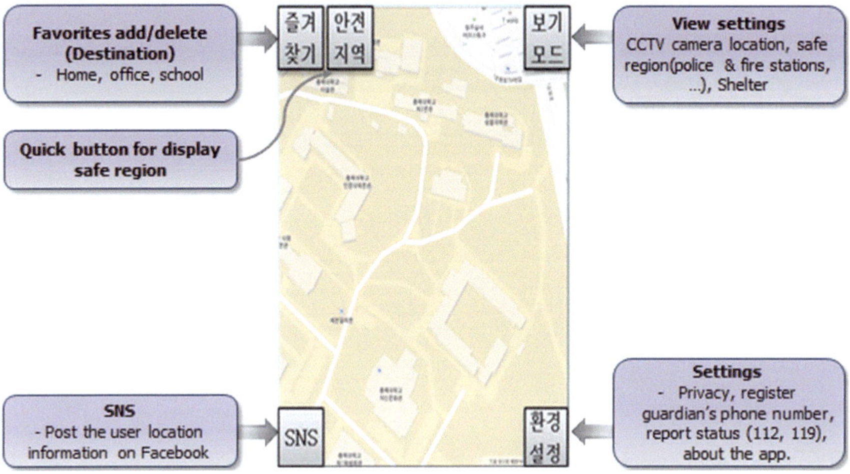

Fig. 9 Composition of the initial screen of the safe-return service

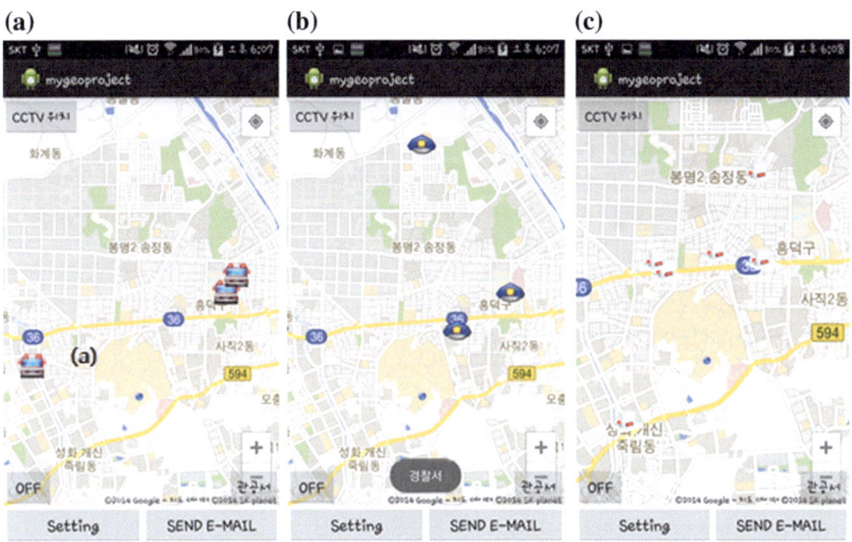

Fig. 10 Visualization examples of safe regions: **a** fire stations, **b** police stations, and **c** hospitals

5 Conclusion

We have described the development of a safe-return service that allows users to be aware of and to prepare for possible dangerous situations in real time. This is accomplished by providing and visualizing information about the possibility of danger through real-time data collection and analysis of various sources of big data regard-

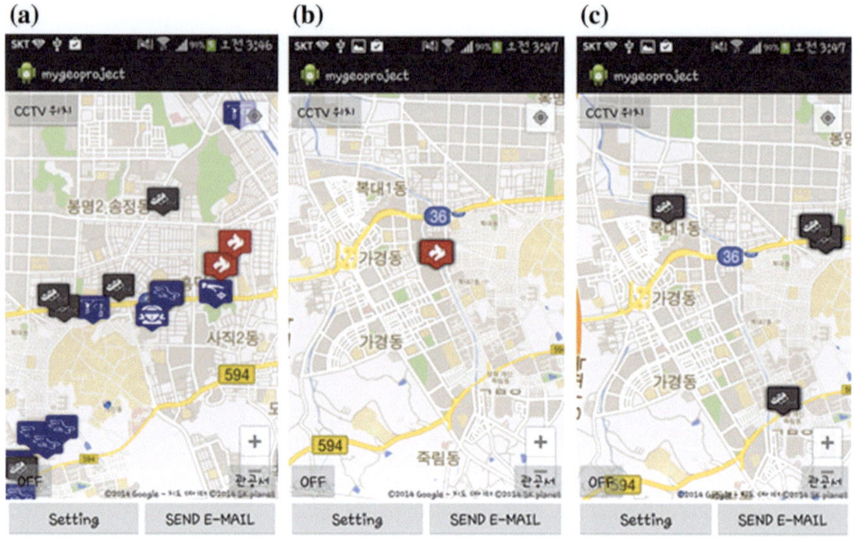

Fig. 11 Visualization examples of dangerous regions: **a** all dangerous situations, **b** disasters (fire), and **c** accidents

ing the user's surroundings. The data that are collected and analyzed include the locations of users, the locations of CCTV cameras, crime/disaster/accident-related real-time news data, the locations of shelters, real-time CCTV video data, and SNS data. We imported the locations of public offices (e.g., police and fire stations, shelters, hospitals) from the Smart Safe-Return system provided by MOSPA, and we used the CCTV video data provided by a portal website. We used online RSS of Chungbuk Ilbo to obtain danger-related real-time data. The collected data were analyzed using Hadoop MapReduce and visualized via Google Maps.

Our safe-return-home service has advantages over existing safe-return services in terms of crime prevention and response. However, we have not yet fully implemented and evaluated the system. First, we designed it to adopt existing crime data; however, the crime data were not reflected in the implemented system because the Korean government has not yet made those data public. Thus, we need to find an alternative way to acquire such data. Second, danger-related real-time data were obtained only from a local newspaper. However, to enhance the reliability of the service, we must obtain such data from the country's major newspapers as well. Third, we need to analyze not only news data but also weather and traffic news to get more accurate results, in addition to reflecting real-time searches on popular web portals.

Acknowledgements This research was financially supported by the MSIT (Ministry of Science and ICT), Korea, under the ITRC (Information Technology Research Center) support program (IITP-2017-2013-0-00881) supervised by the IITP (Institute for Information and Communications Technology Promotion).

References

1. S.-Y. Ju, M.-H. Song, G-A Ryu, M. Kim, K.-H. Yoo, An Embedding Method of Dynamic Educational Content through Big Data Analytic, Advanced Science and Technology Letters, vol. 51 (CEB-CUBE 2014), (2014), pp. 203–206.
2. J. W. Lee, J. K. Kim, H. J. Moon, I. K. Kim, M. H. Son, G. A. Ryu, J. S. Jeong, K. H. Yoo, Development of a Smart Return Service using Big Data Analysis, The 2nd KBDS Conference on "Big Data and Social Safety", (2013), July 03–05 pp. 47–49, Jeju, Korea.
3. Ministry of Security and Public Administration (MOSPA), SOS National Safety Service, http://www.mospa.go.kr/frt/sub/a07/sos/screen.do (accessed on November 2014).
4. Ministry of Security and Public Administration (MOSPA), Smart Safe-Return, http://www.gmap.go.kr/tcportal/csr/CSRSafetyMain.do (accessed on November 2014).
5. Apple Inc., Safe-Return Project, https://itunes.apple.com/kr/app/anjeongwiga/id520573071?mt = 8 (accessed on November 2014).
6. Google Inc., Elegant Homecoming, brothersnsisters.app@gmail.com, https://play.google.com/store/apps/details?id=bns.gracefulreturn&hl=ko (accessed on November 2014).
7. Apple Inc., Taxi Safe U, https://itunes.apple.com/kr/app/taegsiansim-iu/id458589973?mt=8 (accessed on November 2014).
8. Google Inc., Help Me (SOS Emergency Alarm), danalinter@gmail.com, https://play.google.com/store/apps/details?id=com.dain.helpmesos&hl=ko (accessed on November 2014).
9. J.-W. Lee, Design and Implementation of a Safe Return Service based on Big Data, Master Thesis, Chungbuk National University, (2014), South Korea.
10. J.-W. Lee, J.-S. Jeong, M.H. Kim, K.-H. Yoo, Requirements for Developing Safe-Return Home Service based on Bigdata, The proceedings of the FGCN 2014 (2014), December 20–23, Hinan Island.
11. Wikipedia, Apache Hadoop, http://en.wikipedia.org/wiki/Apache_Hadoop (accessed on November 2014).
12. D. Borthakur, J. S. Sarma, J. Gray, Apache Hadoop Goes Realtime at Facebook, Proceedings of the 2011 ACM SIGMOD International Conference on Management of data, (2011), Jung 12–16, pp. 1071–1080, Athens, Greece.
13. C.-T. Chu, S. K. Kim, Y.-A. Lin, Y. Y. Yu, G. Bradski, A. Y. Ng, K. Olukotun, Map-Reduce for Machine Learning on Multicore, Proceedings of the Twentieth Annual Conference on Neural Information Processing Systems, (2006), pp. 281–288, December 04-09, Vancouver, B.C., Canada.
14. KAIST, Hannanum, http://semanticweb.kaist.ac.kr/home/index.php/HanNanum (accessed on November 2014).
15. S.-S. Kang, Y.-T. Kim, A design and implementation of efficient Korean morphological analyzer based on dictionary information, KIISE Spring Annual Conference, (1991), vol. 18, No. 1, pp. 529–532.
16. S.-S Kang, Encoding of Morphological Analysis Result and Eojeol Dictionary Construction, Proceedings of the 16th Conference on Hangul and Korean Language Information Processing, (2004), October 08–09, vol. 16, No. 1, pp. 112–117, Ulsan, Korea.
17. S. Kwak, B. Kim, J. S. Lee, Construction of an efficient Pre-analyzed Dictionary for Korean Morphological Analysis, KIPS Tr. Software and Data Engineering, Vol. 2, No. 13, (2013), pp. 881–888.
18. Wikipedia, tf-idf, http://en.wikipedia.org/wiki/Tf%E2%80%93idf (accessed on November 2014).

Author Index

W. Lee and C. K. Leung (eds.), *Big Data Applications and Services 2017*,
Advances in Intelligent Systems and Computing 770,
https://doi.org/10.1007/978-981-13-0695-2

Subject Index

W. Lee and C. K. Leung (eds.), *Big Data Applications and Services 2017*,
Advances in Intelligent Systems and Computing 770,
https://doi.org/10.1007/978-981-13-0695-2

MIX
Papier aus verantwortungsvollen Quellen
Paper from responsible sources
FSC® C105338

If you have any concerns about our products,
you can contact us on
ProductSafety@springernature.com

In case Publisher is established outside the EU,
the EU authorized representative is:
Springer Nature Customer Service Center GmbH
Europaplatz 3, 69115 Heidelberg, Germany

Printed by Libri Plureos GmbH
in Hamburg, Germany